P9-EDP-487

BIOLOGICAL STUDIES OF
THE
ENGLISH LAKES

Also by T. T. Macan:
A Guide to Freshwater Invertebrate Animals
Freshwater Ecology

1. Buttermere from the air. The main inflow drains the valley from the right

Biological Studies of the English Lakes

T. T. Macan, M.A., Ph.D.

LONGMAN

LONGMAN GROUP LIMITED
London
*Associated companies, branches and representatives
throughout the world*

© *Longman Group Ltd., 1970*

All rights reserved. No part of this publication may be
reproduced, stored in a retrieval system, or transmitted in
any form or by any means, electronic, mechanical, photo-
copying, recording, or otherwise, without the prior
permission of the Copyright owner.

First published 1970

SBN 582 46019 0

*Printed in Great Britain by
The Camelot Press Ltd., London and Southampton*

CAMROSE LUTHERAN COLLEGE
LIBRARY

QH
138
L 35
M 12
16,425

This book is dedicated
with respect and affection
to the memory of

P. A. BUXTON
F. E. FRITSCH
W. H. PEARSALL
J. T. SAUNDERS

Contents

List of Illustrations

List of Plates

Acknowledgements

Thanks are due to Mr. H. C. Gilson, Director of the Freshwater Biological Association, for permission to use photographs of apparatus.

The publishers are grateful to the following for permission to reproduce photographs:
Aerofilms Ltd for Plate 1, and Mr G. V. Berry for Plates 2, 3 and 4.

We are also indebted to the following for permission to use material in the preparation of figures: Blackwell Scientific Publications Ltd for material from *Journal of Animal Ecology, Journal of Ecology* and *New Phytologist*; Cambridge University Press for material from *Proceedings of the Association of Applied Biology*; the Company of Biologists Ltd, Cambridge, and Cambridge University Press for material from *Journal of Experimental Biology*; the Royal Geographical Society; the Linnean Society of London; the Royal Meteorological Society; the Royal Society for material from *Philosophical Transactions*; the Salmon and Trout Association; Springer-Verlag, Berlin, for material from Archives of Microbiology, and for material from Archives of Hydrobiology and the Department of Zoology; University of Washington for material from Limnology and Oceanography.

Preface

Except for part of chapter 10 that which follows is a compilation, and the only reason for writing a book of this kind is that the information is so copious and scattered that the gathering of it together in one place will simplify the task of workers in the future. The writer can be compared to one who industriously picks up stones along a mountain track and piles them to make a cairn that may guide others toiling upwards later. But is it to be just another cairn, one of many that the traveller passes giving it attention which, at best, is brief and fleeting? The writer's vanity demands something more than a mere milestone on the way, something that marks an end and a beginning, that stands out as a principal feature of the route. To that end he looks round for dates, for nothing highlights an event more brightly than celebration of it after the passage of years in certain neat multiples of the decimal system. The present writer has not been unsuccessful in this quest, for the main part of this book was written in 1967 exactly fifty years after W. H. Pearsall, in one of his first papers on the Lake District lakes, laid the foundations of an idea from which much has developed since. It is satisfactory to discover this jubilee and delightful to find that it draws attention to a man whose influence is remembered with so much pleasure.

The purpose of this preface is to express deep gratitude to my colleagues who, once again, have been most generous not only with time spent searching the chapters on their subjects for errors and omissions but also with information that had not been published. Chapters 2 and 12 have been read by Dr Winifred Pennington, chapter 4 by Dr C. H. Mortimer, F.R.S., chapters 5 and 12 by Mr F. J. H. Mackereth, chapter 6 by Dr J. W. G. Lund, F.R.S., chapter 7 by Mr W. J. P. Smyly, chapter 11 by Miss C. Kipling, and chapter 13 by Dr Vera G. Collins.

I offer sincere thanks to my permanent assistants, first Miss Rachel Maudsley (now Mrs Hans Erwig), and then Miss Annette Kitching, for help in the early and late stages respectively of the preparation of the book. To Miss Maudsley I am indebted for many of the original illustrations, to Miss Kitching for help with the proof-reading and the compilation of the index. I also record with gratitude the assistance of Miss Suzanne Hopley and Miss Jill Payne with these later stages.

This preface started by disclosing a jubilee and it is fitting to conclude it with mention of a centenary. It was almost one hundred years ago that Longmans, Green and Co published *Insects at Home* by the Reverend

J. G. Wood, a book that I consulted frequently in my early years as a naturalist. That the same firm might one day publish a book by me seemed then a project too ambitious even to dream about. Now that that has come to pass, I record the pleasure that comes not only from the event but from the collaboration with a firm whose methods are both friendly and businesslike.

CHAPTER 1

Introduction

Ideas creep slowly across the North Sea and the English Channel. The revised calendar was some two hundred years on the way and the metric system will take nearly as long. Scientific ideas are sometimes no exception, which is not always a disadvantage. Few generalizations are wholly true, and not many people would go all the way with Wesenberg Lund (1921), who wrote 'that the study of Nature must always begin with the slightest possible literary ballast'. Nonetheless a problem may be solved more quickly if different workers approach it in a different way, each un-influenced by the ideas of the rest, and the study of the English Lakes provides an example.

Early in the present century Thienemann in Germany and Naumann in Sweden devised a classification of lakes, a scheme that was richly elabor-ated by their disciples in the years that followed. However, as knowledge accumulated, flaws in the system were exposed and eventually, at the inter-national congress of limnology in Finland in 1956, a masterly review by H-J. Elster (1958), with due emphasis on the shortcomings, marked the end of an era. Since then limnologists have been less concerned with forc-ing lakes into categories than with arranging them in a series according to the primary productivity.

Shortly before the First World War broke out, W. H. Pearsall started a study of the English lakes, but neither then nor in later years was he influenced by the ideas that dominated much of the rest of western Europe. He worked on the eleven largest lakes, ignoring Thirlmere, which Man-chester Corporation had acquired and turned into a reservoir during the closing years of the last century. These lakes were all formed during the Ice Age in an area in which there are now no striking climatic or geological differences. There is, however, enough difference in the position of each one relative to the main mountain masses to produce biological differences, and Pearsall could have pointed out that Ennerdale and Wastwater, for example, are typical oligotrophic lakes, whereas Esthwaite is eutrophic according to the continental classification. He did not. He avoided cate-gories and stressed that the lakes formed a series, each one occupying a different place somewhere between the extremes named above. This idea, put forward by Pearsall after his earliest investigations, is similar to the one which was adopted widely elsewhere after the eclipse of the 'trophic' system.

These differences superimposed on fundamental similarity give the English lakes their peculiar interest.

1

The factual side of an introduction must begin with a definition, and then give general information about geography, physiography and climate. Lakes cannot be studied without some knowledge of their history, both

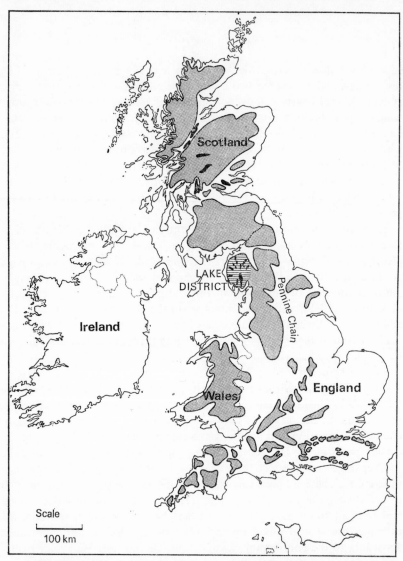

1. The British Isles, showing the position of the Lake District and of high ground in Britain.

geological and recent. Recent history, as will be seen, is particularly important, as man has been the most potent agent of differences between the lakes, which indeed he has been altering since his first arrival in the district. So much of the work on the lakes has been done at the laboratory

of the Freshwater Biological Association that it seems not inappropriate
to include an account of that organization.

The Lake District lies on the west coast of England close to the Scottish

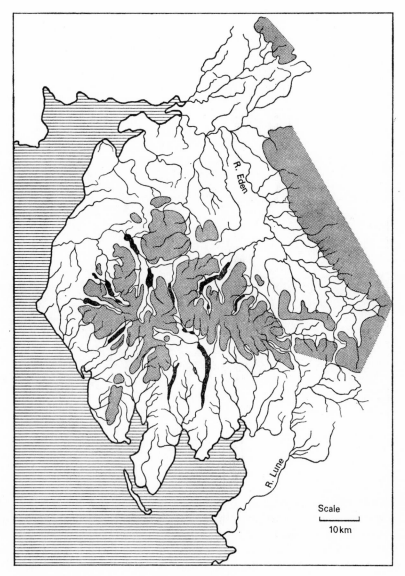

2. Cumbria.

border (fig. 1) within a roughly diamond-shaped projection bounded to
the south-west and north-west by the sea, and to the south-east and north-
east by the valleys of the Rivers Lune and Eden respectively (fig. 2). Some,

3

notably those with an interest in the tourist trade, like to apply the term 'Lake District' to the whole of this area, but here the name of one of the old English kingdoms, Cumbria, is used for it. There are several other definitions, of which the most convenient for biologists is probably that based on the geology by Marr (1916), according to whom the Lake District is the area of old hard rocks that yield little calcium within a ring

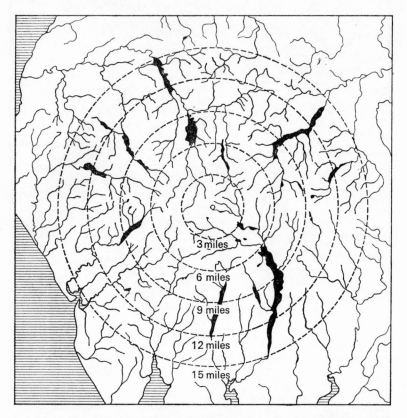

3. The radiate symmetry of the Lake District, shown by the stream lines (Mill, H. R. (1895), *Geogr. J.* **6**).

of the newer calcareous formations. This is a clear boundary, except to the south where islands of older rock break through the limestone, and to the east, where the old rocks are continuous with those of the Pennines (fig. 1). Here an arbitrary boundary must be fixed. Pearsall used the railway and Macan (1940) the main road. Some naturalists hold that it is now more logical to make use of the boundaries of the National Park, which takes in some of the coast on the south-west side and some of the limestone to the south-east.

However defined, the Lake District is small in comparison with similar areas in other countries, and Mill (1895) has pointed out that a circle having a radius of 15 miles (roughly 24 km) encloses all the lakes (fig. 3).

There are three main mountain masses separated by a central valley running north and south and containing Thirlmere, and an east–west valley which drains into Bassenthwaite (fig. 4). The northern bloc is

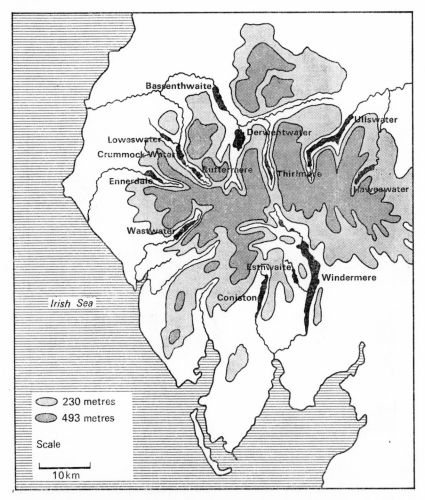

4. Relief map of the Lake District (Pearsall, W. H. and Pennington, W. (1947), *J. Ecol.* **34**).

dominated by Skiddaw, the eastern by Helvellyn, and the western, which is the highest and most rugged, by Sca Fell. All three mountains named, and no others, exceed 3000 feet in altitude, a satisfactory round number in English units. The metric system, opponents to change point out, is not suited to English conditions, as no mountain reaches 1000 m. The highest is but 978 m above sea level, and 3000 feet is 915 m.

Fig. 5 is a simple map designed to introduce the reader to the lakes. The nineteen named have all been studied by biologists of various kinds, and that is the reason for their inclusion. A discussion of exactly what a lake is would be no more profitable than one on the definition of the Lake District. In general the lakes lie in the main valleys and the emergent vegetation is *Phragmites* not *Carex*. Blelham is a lake according to this botanical definition, but it is not in a main valley and is commonly known

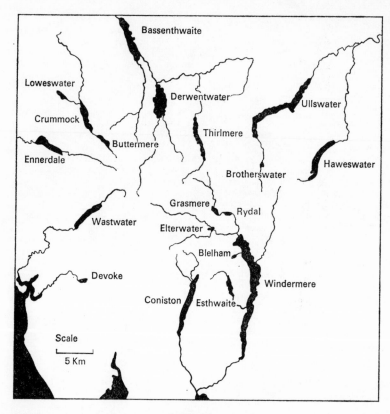

5. The lakes.

as Blelham Tarn, tarn being the name applied to the numerous small bodies of water. There is no emergent vegetation in some of the lakes, which cannot therefore be judged on this criterion. Brothers Water, Devoke Water and Elterwater are small and their exact status is argued about by people so minded.

CLIMATE

There are meteorological stations at Ambleside, just to the north of Windermere, and at Keswick near the north end of Derwentwater (fig. 13).

The local council, who maintain the Ambleside station, kindly send all their records to the Freshwater Biological Association's laboratory at Ferry House. The *Reader's Digest Atlas* (*1965*), which contains useful meteorological information about the whole of the British Isles, quotes the Keswick records. The Freshwater Biological Association itself records rainfall, wind direction and speed, and water temperature at various

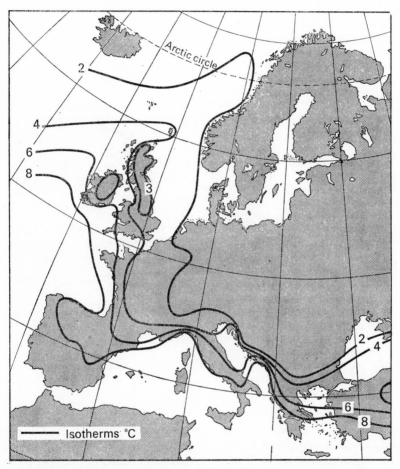

6. January Isotherms (from the Guide to the 12th Congress of Limnology).

points. There are a number of private rain gauges throughout the district.

Windermere lies on the parallel of latitude 50° 30′ N, where the sun is visible for some 17 hours 9 minutes at the summer, and 7 hours 23 minutes at the winter solstice. Superimposed on the seasonal cycle of temperature determined by the sun, is a considerable and irregular variation depending on the origin of the air masses over the country. Those coming off the western and south-western Atlantic bring moist air which is relatively cool in summer and relatively warm in winter, on account of the Gulf Stream, whose influence is shown by the north and south trends of the January isotherms (fig. 6, cf. fig. 7). Air from the opposite direction brings fine clear weather, warm in summer and cold in winter. Polar air is cold at all times of year. At the beginning of June 1953 a polar air stream brought bitterly cold conditions, as is well remembered by those who sat up all night in position to watch the Queen pass to and from her coronation. The maximum temperature at Ambleside on 4 June was 12·2°C and the minimum 6·1°C. On 3 December, six months later, the maximum was the same and the minimum 2·8°C higher.

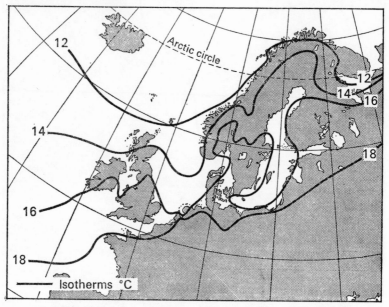

7. July Isotherms (from the Guide to the 12th Congress of Limnology).

Water temperature is discussed more fully in chapter 4 and here a few general remarks will suffice to give an impression of the climate. Summer temperature at Ambleside has exceeded 29·4°C (85°F) in nine years since 1931 when the records start, the highest figure being 31·7°C (89°F) on 9 July 1934. In winter, skating on some of the tarns, for a few days at least, is possible almost every year, provided snow does not fall after the ice has formed. Skating on the smaller lakes, such as Esthwaite and Rydal, is possible perhaps two or three times a decade. Skating took place on Windermere in 1895, 1929 and 1963, and some scientists from the Freshwater Biological Association ventured on to the ice in 1947, to make observations through the snow, which was thick, and the ice. Generally Windermere remains unfrozen; on a lake of this size stillness as well as low temperature is an important factor in ice formation.

Introduction

The winter of 1962–3 was one of the coldest on record. A temperature below freezing point was recorded at Ambleside on 16 November and the minimum was rarely higher until mid-December when there was a week of warmer weather. It was below freezing on 20 December and on Christmas

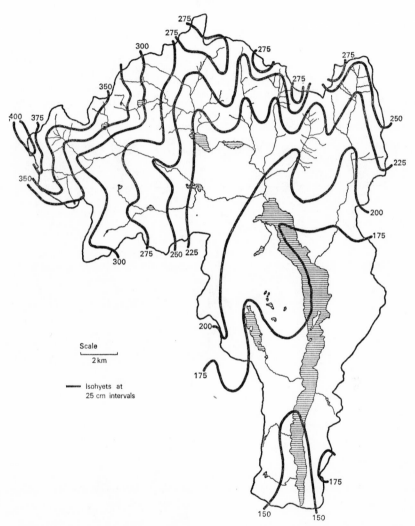

8. Rainfall in the drainage area of Windermere (McClean, W. N. (1940), *Q. Jl R. met. Soc.* **66**).

Day fell to −12·2°C, the coldest night of the winter. The first week in January was warmer with minimum temperatures just above freezing point, and on 11 January started a cold spell which lasted until 6 March. A well known newspaper, the *Manchester Guardian* (now the *Guardian*), published a booklet about the winter, giving a week by week account and some comparisons and analyses. Table 1, a comparison made by the Meteorological Office, is taken from it and the figures for Ambleside have been added.

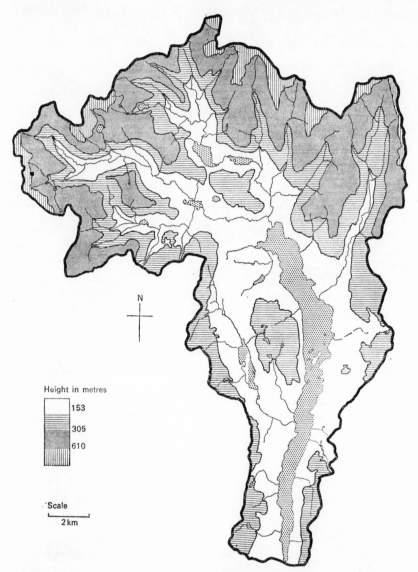

Height in metres

153
305
610

Scale
2 km

9. Topography of the drainage area of Windermere (McClean, W. N. (1940), *Q. Jl. R. met. Soc.* **66**).

The mountains of the Lake District are the highest in England and, as they lie near the western seaboard, they cause it to be the wettest part, though the annual rainfall is greater in the higher mountains of Wales and

Table 1

Comparison of severe winters

A very cold day is one with a maximum temperature of 2·8°C or lower

		No. of con- secutive very cold days	No. of very cold days during winter	No. of days on which tem- perature re- mained below freezing point	Longest run of such days
	1962–3	35	53	18	9
	1946–7	21	42	17	8
	1939–40	12	37	9	3
LONDON	1916–17	20	31	6	2
	1894–5	24	30	11	7
	1890–1	24	42	21	7
	1878–9	16	46	14	5
AMBLESIDE	1962–3	7	50	3	2

Scotland. Fig. 8 shows the rainfall in the Windermere drainage area (McClean 1940) as revealed by the records of twenty-two rain gauges of which all but the two highest, which are read monthly, are read every day. Fig. 9 shows the heights in the same area and from a comparison of the two it is evident that there is a close relationship between altitude and rainfall. The amount falling on the centre in the mountains is well over twice that at the edge and nearly four times the amount falling on the flat coastal plain. Fig. 8 does include the wettest part of the Lake District and seems therefore sufficient for the purpose. A map showing the rainfall in the whole Lake District is published by the Ordnance Survey, but, as it

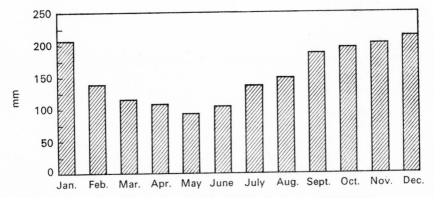

10. Average rainfall per month at Ambleside, for the years 1932–1964.

11

does no more than show the relationship between rainfall and contours and as much of it must be based on fewer gauges than fig. 8, it has not been included.

On average, precipitation is highest in winter, decreases to the middle of the year and then rises again (fig. 10), but in any year there may be wide departure from this pattern. For table 2 some arbitrary figures for low and high monthly totals have been taken. Rainfall may be slight in any month of the year and records of less than 25 mm (approx. 1 inch) are distributed evenly among the months, July and December being the only two excluded and February and May the only two with more than one record. Rainfall between 25 and 50 mm is also fairly evenly distributed though occurring more often in the dry months. The relation between high totals and high averages is closer. February is the least constant month, frequently dry and sometimes very wet, July the most constant, rarely reaching either of the extremes chosen, although it is the month in which the highest daily total was recorded.

A daily total in excess of 75 mm has been recorded eight times, all in the months of October to February inclusive, and over 100 mm once in February and once in July. The highest was 102 mm on 29 July 1938.

The variation in the total rainfall in a year may be seen in fig. 11.

The total number of hours of sunshine at Keswick averaged 1207 during the years 1921–50 (*Reader's Digest Atlas 1965*). The average at

Table 2

Rainfall at Ambleside 1932–1966 inclusive
High and low records. Number of occasions on which the rainfall fell within the limits shown

Under 25 mm	25–50 mm		260–299 mm	300–339 mm	340–379 mm	380–419 mm	420-over 440 mm
1	1	January	4	4	3		
3	5	February	3				
1	4	March	1				
1	4	April					
2	6	May					
1	5	June					
	2	July					
1	2	August	2	1			
1	2	September	2	3	1	1	
1	2	October	3				2
1	1	November	2	5	4	2	
	1	December	5	4	1	3	

Ambleside was 1178 for the years 1932–65. The south and east coasts of England received between 1700 and 1800 hours but Manchester (1071) and Fort William (981) were gloomier (*Reader's Digest Atlas 1965*). Manchester's rainfall is relatively light, some 868 mm per annum, but industrial smog cuts off the sun. Fort William, near the west coast of Scotland, is wetter than Keswick.

It is difficult to make any useful generalization about wind, except that it is generally blowing, because the valleys deflect it. Strength and direction are recorded by means of an anemometer at Ferry House. This probably gives a fairly true reading of the direction of easterly winds,

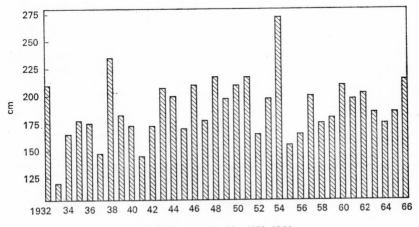

11. Rainfall at Ambleside, 1932–1966.

which often blow for periods during the first half of the year, but, as a range of hills shelters it from the prevailing winds from the west, it records these much less accurately. The typical pattern is an approaching depression causing south or south-westerly winds at first and then westerlies and north-westerlies as it passes. The record on the anemograph chart shows a sudden shift from north to south. This state of affairs persists northwards till the end of the range between Esthwaite and Windermere is reached (fig. 9), and the northernmost half of the north basin is the only part of Windermere where, in a westerly wind, sailing boats commonly find a cross wind. In the south basin of Windermere the hills are lower and a westerly wind is deflected less, but a true cross wind is very rare.

CHAPTER 2

Geology and History

The geological formations that make up the Lake District are complicated in detail but simple in outline, there being three main bands running across it. To the north lie the Skiddaw slates, laid down in early Ordovician times; in the middle lies a wide intrusive mass thrown up by volcanic activity later in the Ordovician period; to the south lie Silurian rocks (fig. 12). Between these and the Borrowdale Volcanic series there is a thin band of limestone, the Coniston limestone. Skiddaw slates are also exposed in the south-west corner, and there is an outcrop of Borrowdale

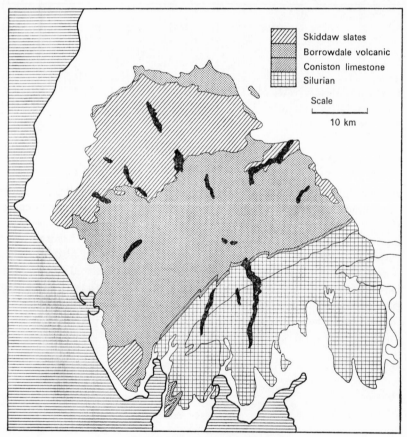

Skiddaw slates
Borrowdale volcanic
Coniston limestone
Silurian

Scale

10 km

12. Geology of the Lake District.

Volcanic rocks in the north. In addition, not shown on fig. 12, there are various later volcanic intrusions scattered throughout the district.

Earth movements at the end of the Silurian folded these rocks into a broad ridge trending ENE/SSW, and then, after a period of denudation, later formations covered them, forming a vast mound. Water running off this carved channels radiating from the centre and gradually cut down to and then into the underlying rocks. Therefore, when these were once more exposed, as a result of denudation, they were dissected by a radial system of drainage bearing little relation to their original folds.

The glaciers of the Ice Age deepened and widened these valleys and carved a basin, later the site of a lake, in almost every one. Moraine material was carried from the centre and deposited in the valleys, mainly at the ends where it sometimes formed a dam which raised the subsequent water level above the edge of the rock basin.

There was considerable rearrangement of morainic material as the ice retreated to higher levels, ice dams burst and water flowing freely became more abundant. Later this process slowed down but there has always been erosion of the mountains and transport of the proceeds into the valleys. The heads of the lakes move slowly down the valleys, and deltas make bulges in the sides which, as is general in glacial lakes, tend to be parallel. Deltas have spread right across two valleys so that two lakes (Crummock and Buttermere; Derwentwater and Bassenthwaite) now occupy the single glacial trench. It is difficult to ascertain how much of this filling process took place in the unstable period of disappearing ice and how much later.

All the Lake District rocks exposed today are hard, and the feature of most importance to the freshwater biologist is that they yield little lime to the waters that flow over them. Not all are poor in lime and when broken some yield appreciable quantities. Calcium, however, is soon leached from a fresh surface, which thereafter gives off very little. For this reason the lakes were richer in calcium and other minerals in their early days than they are now.

All the rocks are hard but some are harder than others. The most rugged relief and spectacular scenery is to be found in the Borrowdale Volcanic series, particularly in the west. The mountains of Skiddaw slate are more rounded and, when broken up by running water or by waves at a lake's edge, give rise to flat stones only a few centimetres across. The other rocks erode into larger fragments, though whether this is important to animals which cling to stones or dwell under them is not known. The shales, slates, grits and flags of the Silurian rocks have also weathered more than the volcanics. Another feature which, as will be seen later, is of importance in the study of lakes is the greater quantity of moraine material near the edge of the Lake District.

The radiating drainage system, explained earlier, is stressed in most books, but like many generalizations it is not completely true, and the exceptions are important to limnologists. The northern mountain mass is cut off from the rest by a river which rises at the edge of the district in an area of clay on which bog occurs and flows inwards into the centrifugal valley in which Derwentwater and Bassenthwaite lie. The drainage area of Loweswater also runs inward, rising almost at the edge of the Lake District and flowing into Crummock Water (fig. 4).

As the present day is approached, the tale of the geologist is taken up by the archaeologist, who, in his turn, gives place to the historian. Biology has contributions to these disciplines, but, in order to avoid all possibility of a circular argument, biological contributions and speculations are omitted from this chapter and brought forward later when the work on cores of mud from the bottoms of lakes is described.

Marr's (1916) account of the geology of the Lake District has been mentioned already; a simpler and more modern account is that of Monk-house (1960).

Knowledge about early man in Cumbria is still fragmentary, and the dating of many of the finds extremely tentative (Rollinson 1967), though there have been important advances since the studies of R. G. Colling-wood (1933) on the archaeology and W. G. Collingwood (1925) on its archaeology and history, for long the standard works. Remains, discovered near the Cumberland coast in recent years, indicate that man reached the district several thousand years before the date given by the Collingwoods. His influence on the landscape was probably negligible.

Neolithic culture, characterized by crops, herds and polished stone axes, is thought to have spread to England from Ireland. The one firm date is that of a Neolithic community near the Cumberland coast, a sample from which was found by means of the radio-carbon technique to be about 5000 years old. The main relics of the Neolithic people are stone circles and axe-heads. The stone circles are most numerous in the south-west but occur along the western and northern fringes of the Lake District and round into the Eden Valley. It is thought that this represents an establishment in the south-west and slow expansion across the more fertile tracts of country, but some stone circles may have been erected by a people who came in from the east. Few circles are situated higher than 200 m.

In 1947 faulty axes, and chips thought to have been produced by the trimming of a stone to make an axe, were discovered near the Langdale Pikes in such quantity as to indicate considerable production. No perfect heads were found, and it is thought that the final grinding and polishing was carried out at the edge of the Lake District where sandstone suitable for the purpose is available. The site is near the centre of the Lake District, and it is clear that there was more traffic into the mountain area than the distribution of other relics suggests. The date, important to the limnologist, at which man took his flocks and herds into the higher regions remains vague. It has been demonstrated that trees can be felled easily with a stone axe.

The influence of the Bronze Age probably came mainly from the east and the two cultures mingled. Little is known about the last millennium B.C. except that towards the end of it the people were building forts in the mountains. There is some diversity of opinion about the identity of the enemy which made this protection necessary.

The Romans built a road through the Lake District with forts at Ambleside, Hardknott and Ravenglass on the coast, and Rollinson brings forward evidence that there was also a fort at Keswick. The main purpose of the penetration is thought to have been a precaution against insurrection by the local tribes. As far as is known the Romans had little influence on the British way of life. When they left, soon after A.D. 400, Angles and Saxons overran England and their place names are now numerous in the Cumbrian plain, from whence presumably the Britons were expelled. From the absence of such names within the Lake District itself it is deduced that the new invaders were not attracted by the steep mountain sides and marshy dales, and that in them life went on much as before.

Late in the ninth or early in the tenth century the Norse, or the Norse-Irish as Rollinson prefers to call them, came to Cumbria. Little definite is known about their coming. Rollinson indeed suggests that some arrived as refugees rather than invaders, having fled from the Isle of Man after what would now be called a unilateral declaration of independence. This provoked the Norse king, who regarded the island as part of his dominions, to plan measures of a severer nature than is currently acceptable on a scale which, when it came to the ears of the islanders, convinced them that the prudent course was to leave quickly and unobtrusively.

Whatever the manner of their coming, the Norse-Irish were soon in occupation of the whole of the Lake District. Nobody chronicled the take-over but the sequence of events has been pieced together from place names and by analogy with the history of Iceland, some of which is on record. They worked their way up the valleys clearing the woods and settling. The main settlement was generally in the lower part of the valleys, higher up there was a summer farm or 'saeter' and beyond pigs were herded in the woods, as is indicated by the location of places that include variations of their word for that animal (Grasmere, Grizedale, Swinside). Collingwood points out the inadvisability of relying too heavily on this kind of evidence, for Swinside could be a corruption of 'Sveyn's saeter'.

Little is known about the fate of the Britons. Possibly they were enslaved and put to work tending the beasts, for a method of counting sheep which is certainly not Norse and appears to have Celtic affinities remained in use long enough to be known today.

The Norse enjoyed an immunity within the Lake District comparable to that of the Britons before them, and their influence is still evident in local speech. The islands and bays of the lakes are 'holms' and 'wykes' respectively, mountains are 'fells', small streams 'becks' and small lakes 'tarns', all words more familiar to Scandinavians than to inhabitants of other parts of the British Isles.

The Norman barons, dividing up England after 1066, did not covet the unproductive fells, and gave them to the church, which built abbeys at various points around the Lake District. The Norse no doubt suffered some loss of status, but a number of them appear to have succeeded in regaining independence fairly soon. The importance of the abbeys to the present theme is their diligence in promoting sheep-farming, which, it must be supposed, led to further removal of woodland to make way for pasture. The monks also started smelting iron ore which led to further pressure on the woods as they were the main source of fuel. In 1565 Queen Elizabeth suppressed all the bloomeries in Furness in order ostensibly to prevent further destruction of the woods, though whether her main concern was for the woods or for the royal exchequer is open to question. The fact is that at the same time a royal company was floated with rights over all the minerals in the Lake District. German miners were imported and some copper was mined. Extraction of this metal and lead has continued ever since though generally with little profit. Quarrying for slate, a thriving industry today, has a history that is possibly continuous back to the Roman occupation.

The life of these Norse farmers, woodcutters, miners and quarrymen went on not greatly affected by events outside, even when these were royal decrees. The middle of the seventeenth century was hard because most of the inhabitants backed the king, who lost the civil war, but with the restoration of the monarchy came prosperity. This declined in the following century as machines came into use and cloth was no longer spun and woven in the country where the wool was grown.

A change of importance to the inhabitants and to present-day biologists started in the middle of the eighteenth century when the wealthy came to look on the Lake District not as a place of wild and fearful desolation but as a picturesque area through which it might be agreeable to travel. Some even settled down, and the presence of writers among them probably speeded the growth in popularity. Southey, the Poet Laureate, lived at Keswick from 1803 till his death in 1843, when the title passed to Wordsworth who had lived most of his life (1770–1850) in the Lake District. These and well known writers of prose advertised the charms of the Lake District, and in 1847 the engineers opened it up to others besides those who could afford to own or hire a carriage by building a railway. This achievement may probably safely be regarded as the beginning of the age of tourism, though numbers in those days were very small compared with those now. It also made it possible for the successful industrialist to set up a home in a part of England that was green and pleasant land but still keep an eye on his dark satanic mills that disfigured the neighbourhood of Manchester or Liverpool. Estates were set up, and trees were planted to improve the view and to provide cover for pheasants; the exclusion of the ubiquitous sheep from gardens probably made regeneration possible too. From the biological point of view, the reforestation may have been the most important aspect of this early stage of the invasion by 'offcomers' as they are called. Later a railway line up to Coniston was built and another right through the Lake District along the valley which isolates the northern mountain bloc (fig. 13). The Windermere line, which runs to London and to the industrial towns of Lancashire, has, however, always been the main route of entry. Incidentally the engineers were unable to bring the railway down to lake level and had to build their terminus near a small village called Birthwaite. Aware that nobody but an inhabitant would wish to buy a ticket to Birthwaite, but that many might be attracted by the well known name of the lake, they named the station Windermere. Hotels, boarding houses, private houses and the shops to serve them quickly sprang up around the station to form a village, which adopted the same name. Now it is general to refer to Windermere and Windermere Lake, and not, in accordance with the priorities, to Windermere Town and Windermere. Here the priority is conceded, tautology is avoided, and 'Windermere' refers to the lake.

Since the Second World War the wages of the lowest-paid workers have risen in relation to the average, and holidays with pay have become more usual. The motor car, formerly owned only by the comparatively well-to-do, is now within the reach of a large proportion of the population. The result has been an enormous increase in the number of visitors to the Lake District. A great many now cóme with tents or in caravans, and most of them spend the night on a regular site where running-water sanitation is compelled by law. The change from an earth latrine to a water closet connected to a sewage works, the effluent of which finds its way ultimately to a lake, is one of importance to the biologist. In recent years it has affected the permanent inhabitant as much as the camper. The writer settled in a small village of 14 houses after the war, having purchased one of the four in which water could be transferred from a well to a cistern in the roof by means of a hand pump. The cistern fed a bath and indoor sanitation, which discharged into a septic tank. The remaining ten houses were provided with earth closets, and the occupants took a bucket to a communal pump when they wanted water. In 1952 mains water was brought to the village, and gradually baths and indoor sanitation were installed in each house. Improvement of this kind has been taking place all the time, in towns as well as small rural communities.

In 1951 the Lake District became a National Park, with boundaries that included not only the Lake District as defined here but a strip of coast to the south-west and some of the limestone area to the south-east. The planning board is charged with the tasks of preserving the natural beauty of the area, of providing access and facilities for outdoor recreation, and of safeguarding the interests of the local communities, notably the farmers. These interests conflict and there is controversy about the widening and improvement of roads, about the use of the lakes to supply water to towns, about forestation and about the siting of campsites, caravan parks and houses. Tents and caravans are kept out of sight as far as possible, and new houses must be in a style in keeping with the existing ones and near them. The permanent population is likely to increase only slowly under this policy, there will be a more rapid increase in the number of temporary dwellers in tents and caravans, and the most rapid increase foreseen is in the number of day visitors as more people come to own cars and as the system of motorways is gradually developed. The main effect on the lakes is likely to be the increased amounts of sewage effluent flowing into them. Whether the increased number of walkers on the fells will increase erosion significantly is more doubtful though already there are scars along the edges of lakes in places easy of access.

Much of this history is relevant not on account of the way events affected the lakes at the time but on account of their contribution to the present pattern of factors operating on the lakes. It is fitting therefore to conclude the historical review with a picture of the Lake District as it is in 1969. Wherever the land is flat it is farmed. The main products are milk, now more profitable than it was, mutton and wool. Sheep graze everywhere, even on the tops of the highest mountains. The most profitable industry, however, is tourism and most tourists come from the south. The Scots to the north and the Yorkshiremen to the west have their own lakes and mountains, and it is the inhabitants of the industrial regions to the south that the Lake District draws in greatest numbers. There is a comparatively narrow approach bounded on one side by the Pennines and on the other by the estuaries of the Lancashire rivers, and travellers by road and by rail are almost bound to pass through Windermere. This town, which is now continuous with Bowness, once a small lakeside village, is the largest inside the Lake District and has developed since the construction of the railway. The development of Ambleside has been similar. It lies on a main road which runs through the centre of the district past Rydal and Grasmere, two other popular holiday and residential resorts (fig. 13). Hawkshead, to the west, was a market town in byegone days when communications were slower, and its importance has declined. Coniston and Glenridding near Patterdale at the head of Ullswater both developed as mining centres. To revert to the route north, the Thirlmere drainage area is almost without habitations, for those that did exist were evacuated and demolished when Manchester Corporation acquired the lake and turned it into a reservoir. Beyond Thirlmere lies Keswick, another old market town, now an important tourist centre. In Borrowdale, as in Langdale, there are numerous hotels and residences. In the remaining valleys the population, both permanent and visiting, is smaller. They are narrower, they are more remote from the lines of communication, and all but the one in which Buttermere and Crummock lie can be entered only from the outside. No road runs over the passes which connect them with the other valleys. Hawes Water has now suffered the same fate as Thirlmere and many of the old habitations now lie beneath the water. In Wasdale and Ennerdale there are a hotel, a youth hostel and a few farms.

The history of the investigation of the lakes may be said to begin in 1895. Before that date random collections of various groups were made but they were not extensive enough to show whether any species now present was then absent, and there is no record of any species disappearing. Their scientific value is, therefore, slight. In the year mentioned H. R. Mill published the results of a bathymetric survey. He worked with a leadline from rowing boats but his experienced geographer's eye enabled him to draw from comparatively scanty soundings contours which recent work with an echo sounder has shown to be remarkably accurate.

Early in the present century W. and G. S. West produced comprehensive lists of the planktonic algae in the lakes. Just before the First World War, and on his return from it, W. H. Pearsall, and his father, who was a local schoolmaster, started work which eventually embraced the physiography, chemistry, rooted plants, phytoplankton and fish of all except the small lakes. It led to the formulation of an idea about the position of each one in a series which has been the basis of a great deal of work ever since. Later workers, brought up to take mechanical propulsion for granted, both on land and on the water, have been equally inspired by the physical achievement. The Pearsalls travelled everywhere on bicycles, often burdened on the homeward journey with a load of water for analysis. On the lakes they rowed.

On his return from the war, Pearsall found a new generation straining to break the narrow conception of morphology and comparative anatomy that had constituted biology up till then. New ideas were circulating, and in this ferment the idea of a laboratory for freshwater research was thrown up. Pearsall (1959) records meeting J. T. Saunders in 1924 and agreeing with him to work to achieve this goal. In 1927 Professor F. E. Fritsch dwelt on the lack of a station of this kind in Britain in his presidential address to the Botanical Section of the British Association for the Advancement of Science. At the same meeting the following year a group of scientists met and formed a committee to approach universities, authorities concerned with the provision of drinking water and the disposal of sewage, angling organizations and private individuals. The response was sufficient to justify the foundation of the Freshwater Biological Association in the following year. It was a bad time for the launching of a new venture, as it was the year in which a severe slump disrupted the economy. The committee's plans had to be reduced drastically and the idea of building a laboratory was abandoned, but they persisted, and in 1931 work started in three rooms of Wray Castle. This was a Victorian folly built beside Windermere in 1840 when both labour and coal were cheap. It was shared with the Youth Hostels Association, and two of the ground floor rooms were open to the public, as were all the grounds. The staff consisted of P. Ullyott and R. S. A. Beauchamp, freshly graduated from Cambridge, assisted by George Thompson who had just left school. Miss P. M. Jenkin, supported by a grant from Newnham College, also took up residence and started work. After a year or two the Youth Hostels Association moved to more suitable quarters, and the Freshwater Biological Association acquired the use of the whole building. Modest increases in staff were found possible, and in 1936 the Development Commission inspected the laboratory and decided that the work justified more Government support than had been forthcoming up till then. Thereafter the Government grant became the main source of funds and superseded the private support which had launched the Association and carried it through the early days. Dr W. H. Pearsall, at that time Reader in Botany at Leeds University, was honorary director during the first few years but the Development Commission's grant was given on condition that there should be a full-time director. E. B. Worthington held that post from 1937 to 1946 when H. C. Gilson was appointed.

After the Second World War there was rapid expansion, and Wray Castle was found to be too small. It was a building that could not have been modified, even if the owners, the National Trust, had permitted it, and in 1950 the laboratory moved four miles down the lake to what had been the Ferry Hotel. The staff at the time of writing (1967) consists of 61 persons of whom 20 have degrees.

Both laboratories having been beside Windermere, much of the work has been done on that lake. It has the disadvantage of being the most infested with trippers, and therefore the lake on which apparatus is most likely to be interfered with. Esthwaite is in private hands, and the owner permits only a few local fishermen to keep boats upon it. Apparatus is, therefore, much safer here and, as it is also close to the laboratory and is the most productive lake in the series, it has been popular for scientific work too. Blelham offers the same advantage. The remaining lakes have been less thoroughly investigated, though various workers have made regular tours to a selection of them in order to obtain comparative data.

The pages that follow are devoted largely to the work of the Freshwater Biological Association but it may be emphasized at this point that that is not the subject of this book. It has been written to give an account of the Lake District lakes, the biotopes they provide and the biocoenoses which inhabit them. It is assumed that every lake has its peculiarities. This approach cuts out work of a more general interest, work that was done at the Windermere laboratory only because that place happened to offer convenient facilities, and which would have yielded similar results anywhere else. Clearly it would be inappropriate in a work of this kind to make more than passing mention of the extensive taxonomic studies that have been made by both botanists and zoologists. Such studies are an essential preliminary to field investigations. When these have reached a certain stage—the word 'completed' is to be avoided in this context—the next step is experimental. How much of this work to include is less easy to decide. If it explains why a certain species occurs in one of the Lake District lakes and not in another, it is undoubtedly relevant. If it explains why a species is found in lakes but not in ponds, it is of less relevance. The line, however, is hard to draw and the drawing of it has led to difficult decisions. It did seem, however, that the work of such colleagues as G. Fryer, D. R. Swift and J. F. Talling contributed to limnology in general rather than to an understanding of the Lake District lakes in particular. Fryer's main work is on the morphology and method of feeding of Cladocera; Swift studied the physiology of fish, mainly trout; Talling is engaged on an investigation of the relationship between phytoplankton and light.

It would be easy to write at length of the men who founded the Association and tended it until it could stand on its own feet. Reginald Beddington, president from 1932 to 1950, comes to mind as a generous benefactor and a valuable link with the angling world. Alderman Sir Albert Atkey, treasurer from 1933 to 1947, was a self-made man, who, having made a success of a business career, turned to local government and became specially interested in water supplies. His liaison with that sphere of activity helped the Association greatly. There were many others. Here, however, I wish to pay tribute to the four scientists to whom this book is dedicated. To them fell the task of watching over the Association in its early days when the research workers were very young and the financial situation uncertain. That progress continued steadily to a point where the Association became an accepted feature of the scientific scene was due to their wise and conscientious guidance. It is not a complete appreciation that is attempted. I know nothing of their deliberations in committee and doubt whether a study of minutes written thirty years ago would add anything to personal impressions. Their interest continued till their deaths but after the war the Association was travelling steadily under its own momentum with its own director. I shall write of them here as they appeared to a young scientist in the early days when their influence was critical.

All four had much in common. They all held high posts in the University world; they were prepared to devote a lot of time, in spite of the exacting requirements of the posts they held, to the interests of the Association; there was never the slightest feeling that thoughts of personal advantage or self-advancement inspired their actions; their visits, after some initial awe had worn off, were sources of pleasure to the staff at Wray Castle. Each one also had something special to contribute.

F. E. Fritsch, Professor of Botany at Queen Mary College, London, had a formidable reputation as an algologist and the wide distribution of his students testified to his success as a teacher. In the chair at a meeting he had the enviable ability to keep speakers to the point in the most good-humoured and friendly way. As he regarded himself as a specialist, and as age and indifferent health prevented him taking part in field excursions and recreational activities, his contacts with those who were not algologists were slighter than those of the other three. On the other hand he was more gregarious than they were, which, together with a good command of German, was probably the reason why he was the only one to belong to the International Association of Limnology, of which he was a founder member. It was he who introduced the young members of the staff to the congresses of this organization and thereby prevented them developing an excessively insular attitude, as could easily have happened otherwise. That absence of contact may sometimes be beneficial to the development of new ideas has already been suggested, but too long an absence can only be disadvantageous. When the congress was held in Britain in 1953, Fritsch was elected President. As the International Association of Limnology had been founded by Germans and had been run by them until the war, and as memories of the war were still clear, the choice of a president from one of the major belligerent countries was an outstanding tribute to his personality. To everybody's grief he died in the following spring.

J. T. Saunders, returning to Cambridge after service in the First World War, became interested in the measurement of pH and temperature, but his contributions to research, though notable, were not large, as he turned gradually to university administration. He was appointed tutor of Christ's, at which college the writer had the good fortune to be a student. In 1934 he became Secretary-General of the Faculties, and in 1953 left Cambridge for the post of principal of Ibadan University in Nigeria. This administrative talent was invaluable to the Freshwater Biological Association. Saunders was the man who, holding his own in scientific discussion, could see at the same time all the details which any plan would involve and the measures that would have to be taken. Typically it was he who foresaw the day when married members of the staff would want houses near the laboratory and was instrumental in purchasing a few. He was very fond of children and there was possibly something parental in his interest in the Association. Like a good parent he watched carefully but said little and was content to remain ready to intervene only when the price of any experience appeared likely to be too high, a situation which, as far as I know, never arose. He was one of the most frequent visitors to the Castle, often bringing his family to lodgings in Wray village for the holidays. A bystander might have seen in him no more than a don relaxing from an arduous task in congenial surroundings. He enjoyed a sail on the lake and liked nothing better than to take people out in his car, a good one, either for business or pleasure. One evening he took everybody down to Black-pool to see the lights. However, appearances were deceptive, and all the while Saunders was keeping himself familiar with every detail of what was going on. He did some work on and with a thermocouple method of measuring temperature, but gradually administration came to take up all his time. He had wide knowledge of freshwater biology and a discussion with him was always productive. His main mark on the subject outside the Freshwater Biological Association was as a teacher and he could number among his pupils both directors of the Association and the first two naturalists; F. T. K. Pentelow whose work on rivers was one of the early British contributions to the subject; B. A. Southgate who, starting as one of the surveyors of the Tees turned to the applied side and ended as director of the Water Pollution Research Laboratory; L. C. Beadle, C. F. Hickling and G. E. Hutchinson to mention three who made a name for themselves further afield. In those days Cambridge was almost the only university where freshwater biology was taught. H. C. Gilson continued the course but, when he came to Wray Castle, it became almost the only university where there was no instruction in this subject.

W. H. Pearsall, Professor of Botany first at Sheffield and then at University College, London, in his later years, also appeared to enjoy his visits to Wray Castle. A keen fisherman and an accomplished artist, he lived life with great zest. After the staff increased in 1935, work was in progress on physiology of invertebrates, fish, ecology of invertebrates, chemistry and algae. Pearsall used to come over, or occasionally we went to Leeds, and devote an hour or so to discussing the work in all these fields. As a result the way through or round many a research impasse became clear, and each worker was left with ideas enough to keep him or her busy for a long time to come. A detached lay observer watching Pearsall would have gained no inkling that it was unusual for a botanist to be capable of discussing constructively five lines of research in three disciplines. A real teacher, he was never more pleased than when he found that somebody had thought for himself and discovered flaws in any of his suggestions. He continued to bubble over with theories and speculations and retained very wide interests till the end of his days. The older members of the staff, at least, retained the habit of making a mental note to ask Pearsall about it when-ever a scientific issue was doubtful or unclear. It was a sad shock when we learned that there would be no more discussions with Pearsall.

P. A. Buxton was working on lice in the years before the war, and, kept in place on his leg with his garter, was a box of them covered with gauze through which they could feed. Periodically he would remove the box to scratch, which sometimes caused surprise when not everybody in the company was a scientist. Had anybody commented he would probably have replied that he was engaged on important work and saw no reason to sacrifice his personal comfort to any convention. It might be said to have been a hobby of his to let air into stuffy places and to throw down a challenge to minds that were passing through life blinkered by convention. His habit of brushing his hair forward in a fringe and of wearing gold-rimmed spectacles of the narrow oval shape fashionable in the eighteenth century was a perpetual challenge. Not everybody took to him on first acquaintance as a result, but those who came to know him soon discovered him to be wise, entertaining and very kind. When he came across a reference which he thought that somebody at Wray Castle might have missed but ought to know about he would scribble it on a postcard, a gesture from a busy professor that was much appreciated. When, during the war, I arrived at an overseas destination that was supposed to be secret (though in fact the Intelligence Section had forgotten to tell the Army Post Office not to stamp letters with the place of origin) Buxton, who had inside information, sat down to write a long letter to my wife not only to give news about my safe arrival but to tell her all about the country, which was one in which he had served during the previous war.

Buxton was Professor of Entomology at the London School of Hygiene and Tropical Medicine, an arduous post because time was taken up with a stream of visitors from all over the world as well as by the usual duties of a head of department. He was, however, a great expert in fitting the maximum amount of work into a busy day. It was characteristic of him to take up the study of fungicolous Diptera towards the end of his life, in order to keep in touch with general entomology and to have something to do during his retirement, the retirement which he did not live long enough to see. He had worked on *Glossina* before *Pediculus* neither of which groups are aquatic. Nor are the animals of deserts on which he had written a book based on observations made during his first-war service, a book which incidentally was of such merit as to be deemed worth reprinting after the second war. He had, however, carried out important work on mosquitoes in Polynesia and Melanesia, and during the course of it became aware of the importance to a malariologist of a good knowledge of ecology. It was this that prompted him to take an interest in the Freshwater Biological Association and to find time to pay regular visits to Wray Castle. Here he too readily took part in whatever was going on. Enquiring one Sunday morning what was planned for the day, he learnt that one of the staff had cajoled his colleagues into carrying picks and shovels up to the top of a mountain, there to dig an experimental pond. Buxton shouldered his tool and climbed and dug with the best.

If he was interested to keep medical entomologists in close touch with the work of academic ecologists, he was also concerned to prevent the Freshwater Biological Association becoming an academic ivory tower, out of touch with those engaged on practical problems. His siege-train for ivory towers, a mixture of genial iconoclasm, learning and charm, was a formidable weapon.

It is to be hoped that as these men looked at the Freshwater Biological Association in their later years and compared it with what it had been at the beginning, they recalled Sir Christopher Wren's words:

Si monumentem requiris circumspice

CHAPTER 3

The Lakes Today

MORPHOMETRY

Pearsall's arrangement of the lakes in a series has been alluded to already. At one end come Wastwater, Ennerdale and Buttermere, which he called 'primitive' or 'rocky', at the other the 'evolved' or 'silted' lakes Windermere and Esthwaite, with the rest strung out in between. The original ideas have been modified as a result of subsequent work, and terms preferred today are 'unproductive' and 'productive', but in essence they remain unchanged and have provided a basis for many later investigations.

Fig. 4 shows the position of the lakes in relation to two contours chosen by Pearsall (Pearsall and Pennington 1947). Wastwater, Ennerdale and Buttermere (plate 1) lie within the main mountain mass, with some of the highest peaks towering at the heads of their valleys. Windermere and Esthwaite (plate 2) lie further from the centre in terrain with a gentler relief. Rain falling on the drainage areas of the first runs over much bare rock and scree, and enters the lake with little in solution. Rain falling on the drainage areas of the second percolates through soil derived from the weathering of the softer rocks and from the moraines left round the periphery of the mountain masses, and reaches the lake with a higher concentration of dissolved substances. This is the primary difference. A secondary difference is due to the small area suitable for cultivation and human settlement in the steep-sided valleys where the unproductive lakes lie. More of the drainage area of the productive lakes is flat enough for these purposes, and therefore the water draining into the lake brings in more of the fertilizers that the farmer has spread on his land and of the sewage, or the products of its decomposition, that emanates from the human settlements.

Settlement is also influenced by the lines of communication, as described in the previous chapter, and it happens that the lakes in whose drainage areas most settlement is possible are also the lakes nearest the main route of entry from the south (fig. 13). The two influences, therefore, reinforce one another. Had the Lake District lain at right angles to its present axis, so that Wasdale faced south, and lay nearest the big centres of population, it would no doubt have been more heavily colonized. Then on the fundamental lack of productivity due to its steep and rocky nature would have been super-imposed a secondary richness due to its position.

Two of the first criteria used by Pearsall were the percentage of the drainage area cultivable and the percentage of the lake bottom to a depth of 30 feet (9 m) rocky. The first depends on a subjective judgement, but, as the judgement was all done by one man and he an astute and practical observer, it can be relied on. The second was obtained by the two Pearsalls

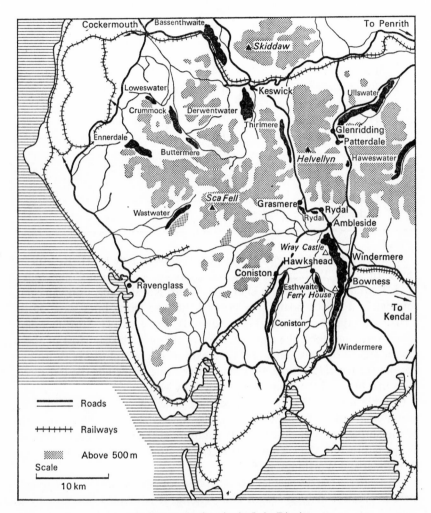

13. Communications in the Lake District.

rowing round the edges of the lakes with a lead and recording whether it struck hard or soft bottom at each sounding (table 3). Other criteria, some due to Pearsall, others to later workers, are recorded in the appropriate place in later chapters.

It will be noted from table 3 that the Pearsalls did not study the lakes smaller than Buttermere, or Thirlmere which had been converted into a reservoir at the end of the previous century.

29

Table 3
The Lake Series (Pearsall 1921)

Lake	% drainage area cultivable	% of lake bottom to depth of 9 m rocky
Wastwater	5·2	73
Ennerdale	5·4	66
Buttermere	6·0	50
Crummock	8·0	47
Haweswater	7·7	25
Derwentwater	10·0	33
Bassenthwaite	29·4	29
Coniston	21·8	27
Windermere	29·4	28
Ullswater	16·6	28
Esthwaite	45·4	12

A bathymetrical survey of the larger lakes was made by H. R. Mill in 1895. In 1937 Windermere was surveyed by Lieutenant-Commander W. I. Farquharson and Petty Officer B. Ward, Royal Navy, using an echo sounder in a motor boat. They crossed the lake along lines some 55 m apart. After the completion of this Admiralty survey, C. H. Mortimer and some colleagues took the echo sounder to the other lakes. Most of their data is still in the raw state, and I am extremely grateful to Mr E. Ramsbottom for the calculations that made it possible to include the last three lakes in table 4, all small ones not surveyed by Mill.

In table 4 the lakes are in descending order of volume, in table 5 in descending order of greatest depth, the figures showing the percentage of the surface area enclosed by the contours at the head of each column. Table 6 presents data about Windermere from the Admiralty survey (Mortimer and Worthington 1942).

Wastwater is the deepest lake though it is not particularly long. Windermere is greatest in both length and volume. It is divided into two by a shallow area beset with rocky islands, and biological work has shown that the two basins may be regarded as two lakes. Both Derwentwater and Bassenthwaite are wide and shallow. Mill takes analysis along these lines further, but, as biological differences have not yet been related to the different shapes, these figures may be left, for the moment, without any more comment.

Table 7 compares the results of the two surveys of Windermere. Five million m³ must be deducted from Mill's figures for volume to allow for a higher lake level, after which the difference between the two is 4 per cent.

Table 4

Morphometry of the English Lakes (data from Mill, H. R. 1895. *Geogr. J.* 6 except those for last three)

Name	Length kiloms	Breadth, metres		Depth, metres		Height of lake surface above sea, metres	Area, sq. kiloms	Volume, million cubic metres	Total drainage area, sq. kiloms
		Max.	Avge.	Max.	Avge.				
Windermere	17·0	1475	869	66·8	23·8	39·6	14·79	347·0	230·5
Ullswater	11·8	1005	756	62·5	25·3	145·0	8·94	223·0	145·5
Wastwater	4·8	805	595	78·6	41·0	61·0	2·91	117·0	48·5
Coniston	8·7	795	549	56·1	24·1	43·5	4·91	113·3	60·7
Crummock	4·0	914	640	43·9	26·7	98·0	2·52	66·4	43·6
Ennerdale	3·8	914	732	45·1	18·9	112·5	2·91	56·0	44·1
Bassenthwaite	6·2	1190	869	21·3	5·5	68·0	5·35	29·0	237·9†
Derwentwater	4·6	1950	1160	22·0	5·5	74·5	5·35	29·0	82·7
Haweswater*	3·7	549	370	31·4	12·0	211·5	1·40	16·7	29·1
Buttermere	2·0	613	567	28·6	16·6	101·0	0·94	15·2	16·9
Esthwaite	2·5	604	405	15·5	6·4	65·5	1·00	6·4	14·0
Loweswater	1·8	539	357	16·0	8·4	120·0	0·64	5·4	6·5
Blelham	0·8	317	133	15·1	6·6	44·2	0·11	0·7	2·3

* Before it became a reservoir. † Does not include the drainage areas of Derwentwater and Thirlmere.

Table 5

English Lakes. Percentage of superficial area covered by different depths of water. (Source as table 4).

Lake	Metres (approx.)							Over 75
	0–3	3–7·5	7·5–15	15–30	30–45	45–60	60–75	75
Derwentwater	35·4	45·0	13·5	6·1	–	–	–	–
Bassenthwaite	43·2	30·8	20·1	5·9	–	–	–	–
Buttermere	22·6		15·2	62·2	–	–	–	–
Crummock	16·0		11·0	20·0	53·0	–	–	–
Haweswater*	41·9		22·1	34·3	1·7	–	–	–
Ennerdale	39·8		12·8	15·2	32·2	–	–	–
Coniston	25·9		11·2	23·7	30·0	9·2	–	–
Ullswater	17·6		17·6	28·8	18·6	15·8	1·6	–
Windermere	42·0			21·6	23·0	10·4	3·0	–
Wastwater	23·0			16·0	16·6	14·1	16·6	13·7

Esthwaite	0–2	2–5	5–8	8–10	10–12	12–14	14–15·5	
	25	20	15	12	17	9	2	
Loweswater	0–5	5–10	10–15	15–16	–	–	–	–
	34	25	33	8	–	–	–	–
Blelham	0–2	2–4	4–6	6–8	8–10	10–12	12–14	14–15·1
	19	13	18	13	11	14	8	4

* Before it became a reservoir.

33

Table 6

Morphometric data for Windermere

Figures in brackets are percentages (Mortimer, C. H. and Worthington, E. B. 1942. *J. Anim. Ecol.* **11**)

Con-tour m	Area enclosed by contour km²*			Layer m	Volume of layer million m³‡		
	North basin	South basin	Whole lake		North basin	South basin	Whole lake
0	8·16 (100)	6·66 (100)	14·82 (100)	0– 2	15·3 (7·2)	12·3 (10·5)	27·6 (8·4)
2	7·11 (87)	5·69 (85)	12·80 (86)	2– 5	20·0 (9·4)	15·9 (13·5)	35·9 (10·9)
5	6·21 (76)	4·90 (74)	11·11 (75)	5–10	29·3 (13·8)	22·2 (18·8)	51·5 (15·6)
10	5·50 (67)	3·99 (60)	9·49 (64)	10–15	26·1 (12·3)	17·9 (15·2)	44·0 (13·3)
15	4·95 (61)	3·17 (48)	8·12 (55)	15–20	23·5 (11·1)	14·7 (12·5)	38·2 (11·6)
20	4·45 (55)	2·72 (41)	7·17 (48)	20–25	20·8 (9·8)	12·7 (10·8)	33·5 (10·1)
25	3·88 (48)	2·36 (35)	6·24 (42)	25–30	18·2 (8·6)	10·4 (8·8)	28·6 (8·7)
30	3·40 (42)	1·81 (27)	5·21 (35)	30–35	16·0 (7·6)	7·50 (6·4)	23·5 (7·1)
35	3·00 (37)	1·19 (18)	4·19 (28)	35–40	14·0 (6·6)	3·60 (3·1)	17·6 (5·3)
40	2·60 (32)	0·25 (3·8)	2·85 (19)	40–45	11·3 (5·3)	†0·50 (0·4)	11·8 (3·6)
45	1·92 (24)	–	1·92 (13)	45–50	7·88 (3·7)	–	7·88 (2·4)
50	1·23 (15)	–	1·23 (8·3)	50–55	5·20 (2·5)	–	5·20 (1·6)
55	0·85 (10)	–	0·85 (5·7)	55–60	3·25 (1·5)	–	3·25 (1·0)
60	0·45 (5·5)	–	0·45 (3·0)	60–65	1·32 (0·6)	–	1·32 (0·4)
65	0·08 (1·0)	–	0·08 (0·5)	65–67	0·08 (0·04)	–	0·08 (0·02)
				Totals	212 (100·0)	118 (100·0)	330 (100·0)

* × 0·384 = mile². † 40–44 m.

‡ × 220 = million gal.; × 35·2 = million ft³. Mortimer has since communicated personally that the echo soundings were too deep by 4%.

Table 7

Comparison of the results of the surveys of Mill and the Admiralty
(Mortimer, C. H. and Worthington, E. B. 1942 *J. Anim. Ecol* **11**)

	Mill	Admiralty
volume m³ × 10⁶	347	330
area km²	14·79	14·82
mean depth m	23·8	22·3

LEVEL OF WINDERMERE

The level of Windermere has been recorded daily since 1933 by staff of the Freshwater Biological Association. The original Ordnance Survey of the lake was evidently carried out during a spell of fine weather, as the level is recorded as being 128·0 feet (39·01 m) above sea level. Mortimer and Worthington (1942) take mean lake level as 129·0 feet (39·32 m). Actually the lake falls to the lower figure or near it fairly soon during a spell of rainless weather and 128 feet (39 m) is, therefore, a more convenient base line for biological purposes. Table 8 shows the extremes of lake level recorded by the Freshwater Biological Association. The total range is 6·78 feet (2·07 m) but in most years it is about half this. Table 9 gives figures for what was probably the most spectacular rise and fall during the period of observation. After the comparatively low level of 129·02 on 28 August 1938 the level was 4·26 feet (1·3 m) higher three days later as a result of the heaviest rainfall recorded during one day (p. 12). Fine weather followed, and the level was back at the starting point after ten days, the rate of fall decreasing rapidly. Whether the reading of 9 August is an error or whether the check in the rate of fall is due to rain which was

Table 8

Levels of Windermere

Extremes recorded during the years 1933–1966

	m	ft		
Lowest level	38·92	127·70	1–3 July	1940
Highest low level in one year	39·11	128·31	10 Aug	1944
Lowest high level in one year	39·94	131·03	17 Dec	1956
Highest level	40·99	134·48	3 Dec	1954

Table 9

Levels of Windermere

Rise and fall and the wettest day recorded in the period 1931–1966

Date	28	29	30	31	1	2	3	4	5	6	7	8	9	10
Level	39·33 m (129·02 ft)	–	–	40·62 m (133·28 ft)	–	–	–	–	–	–	–	–	–	39·33 m (129·02 ft)
Difference compared with previous day	–	+82	+735	+482	–223	–232	–210	–143	–128	–95	–61	–61	–12	–104 mm
Rainfall at Ambleside	8·1	102·1	23·9	tr	0	0	0	0	0·25	tr	4·6	tr	0	0 mm

35

heavy further up the valleys but very slight at Ambleside, the recording station, cannot now be determined. Mill (1895) records a few observations on other lakes, from which it seems likely that none is greatly different. Derwentwater has a slightly larger range. Fig. 14, kindly provided by Mr R. L. Harrison, Chief Engineer, shows the totally different régime in a lake that has become a reservoir.

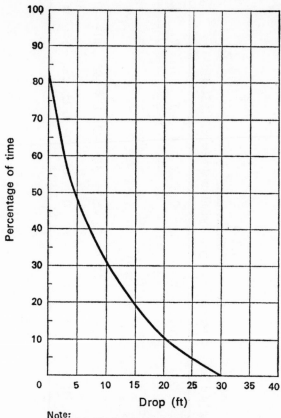

Note:
At overflow level for 17% of time
Below 10' below cill level for 31% of time
Below 15' below cill level for 18% of time
Below 20' below cill level for 10% of time

14. Levels of Thirlmere, 1941–1966, showing the percentage of time when the lake was at, or below, a given level (information supplied by Mr R. L. Harrison, Manchester Corporation Waterworks).

THE LAKE SHORES

Investigations in the early days by Moon (1934) were based on Wray Castle, which stands in a commanding position on an eminence of rock. This descends steeply through Watbarrow Wood to a shore of rocky cliffs,

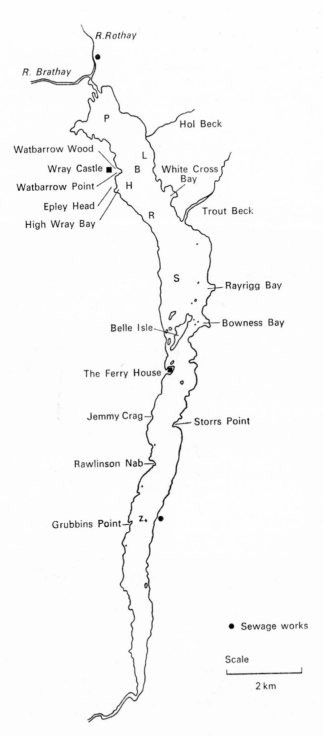

R.Rothay

R. Brathay

Hol Beck

P

Watbarrow Wood

Wray Castle ■   B

L

White Cross
Bay

Watbarrow Point

H

Epley Head

R

Trout Beck

High Wray Bay

S

Rayrigg Bay

Belle Isle

Bowness Bay

The Ferry House

Jemmy Crag

Storrs Point

Rawlinson Nab

Grubbins Point

Z

● Sewage works

Scale

2 km

15. Windermere, showing features mentioned in the text.

of which the most conspicuous feature is Watbarrow Point (fig. 15). Running out into the lake, this point presents on its southern exposed side a smooth flat surface descending steeply and far into the water. To the south there has been disintegration under the battering of the waves and there is a horizontal shelf covered by flat stones of all sizes, the products of the erosion. Moon calls this the Bannisdale shore, but here, in a more

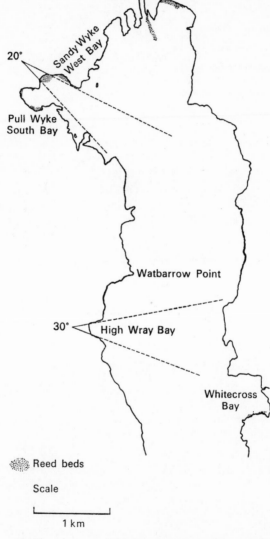

16. The north end of Windermere showing the reed-beds (Macan, T. T. and Worthington, E. B. (1951), *Life in Lakes and Rivers*).

comprehensive study that includes rocks that are not Bannisdale slate, the term 'rock' shore is applied.

Adjoining Watbarrow Wood to the south is a field in which lies Epley Head, a large mound of morainic material also known as boulder clay or glacial drift, whence the term 'drift' shore applied by Moon to this substratum. The relative positions of woodland and pasture, here and elsewhere, are not fortuitous or due to the whim of some bygone cultivator. Where there is little drift covering the underlying rocks, and outcrops of the latter are so frequent that the land is not worth the trouble of clearing and cultivating, the woods have been left. Where the drift is thicker, the trees have been removed and the area has been laid down to grass. Waves beating upon glacial drift have washed away the clay and larger particles and left behind the stones and boulders embedded in the moraine (fig. 17). These vary in size from occasional huge boulders down to pebbles a few centimetres across. All are more rounded than those on the rock shore

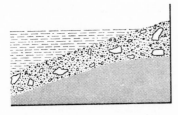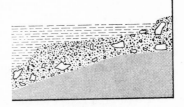

17. The erosion of a boulder clay shore, giving a wave-cut platform (Macan, T. T. and Worthington, E. B. (1951), *Life in Lakes and Rivers*).

owing to the abrasion suffered during transportation by the ice. When erosion has removed the fine particles and left a substratum of stones and boulders, these afford protection to the finer material below them and hinder further removal. There is, therefore, in places, sufficient soil for a sparse growth of rooted plants, of which the chief is *Littorella uniflora*. At the south end of the stretch of drift shore under discussion is an inlet, High Wray Bay, sufficiently deeply indented to enjoy a measure of shelter from gales, except those blowing from a comparatively narrow sector (fig. 16). In consequence it is floored with sand, but it is not sheltered enough for extensive plant growth. In several bays which are more protected there are beds of *Phragmites communis* and often a thick sward of *Littorella* as well.

Wave action is also reduced with increasing depth, and consequently the products of the erosion of a moraine shore tend to settle in a graded series, the coarsest in the shallow water and the finest further out (fig. 17).

Moon chose this part of the lake for his work because it lay nearest to the laboratory. By chance he had selected one of the most imposing outcrops of rock and one of the largest deposits of moraine material to be found anywhere round the lake. Most of the remaining shores are similar but the distinction is frequently less clear, pockets of drift lying between ribs of rock, with the result that the substratum of the lake is derived from both.

The rivers Rothay and Brathay, which join some 500 yards higher up, enter Windermere beside two rocky promontories, and all the land to the west is rocky. Delta extends only eastward from the mouth of the river. The deltas of most other lakes stretch from one side of the valley to the other. Windermere's delta is small because the lake lies far down the valley, and the material washed from the mountains is not carried beyond the long flat stretches that lie above the lake. The River Rothay runs through Grasmere and Rydal, and the River Brathay through Elterwater before reaching Windermere (fig. 5).

The Troutbeck, in contrast, enters the lake at the south end of a large deltaic plain. This contains few large stones where it bounds the lake, which is here floored with sand and gravel and some small stones. The lake was formerly shallow for a considerable distance and the bottom was carpeted with vegetation sufficiently thick to warrant the name 'The *Littorella* Sward', but in recent years it has been extensively excavated as a source of sand and gravel for commercial purposes.

The other lakes are similar though several have peculiar features that must be mentioned. Blelham, a small piece of water (table 4) is completely fringed by reed-beds except off one small rocky point. Esthwaite, another small lake (table 4), has a complete fringe along the west side, but a stony substratum along the other, which is exposed to the prevailing wind. In the unproductive lakes there are no reed-beds. Wastwater is noted for its screes which run down into the water along nearly the whole of the south side. The Skiddaw slates, as mentioned earlier, tend to break into smaller pieces than the other rocks. The north-west corner of Bassenthwaite is bounded by a delta which produces a substratum similar to that of the *Littorella* sward in Windermere. The small size of the present stream, and the uniform nature of the material, suggest that it may be a glacial outwash fan that was deposited during some cataclysm as the ice was leaving the Lake District. The Troutbeck Delta was probably formed in the same way.

A description of the countryside in which the lakes lie, though relevant to this chapter, seemed to fall more naturally into the one before, as a culmination to the account of the history. Population, on the other hand, seems to be more pertinent here among the details about the lakes.

What the limnologist wants to know is how much sewage, purified or not, enters each lake. This depends first on the total population, which must be divided into groups according to the ease with which numbers can be counted. There are permanent residents, of whom a record is kept for taxation and voting purposes, day visitors, and those who, coming for longer holidays, must be divided further into dwellers in houses and dwellers in tents or caravans. The sewage load of a lake depends secondly on the method of sewage disposal. Where earth closets still exist, there is likely to be little effect on the lake; the soil also absorbs most of the minerals in the effluent from a well sited septic tank; but sewage subjected to more modern methods of treatment is in the lake quite soon, having lost little of its fertilizing capacity on the way. Unfortunately information about the number of visitors or about the relative importance of the various methods of sewage disposal is scanty; perhaps one day a limnologist will decide that this is a line which does come within his or her field. In the meantime there is nothing to be done except to present such figures as there are and to discuss their usefulness in the present context.

It is easy to calculate the population in a drainage area such as that of Ennerdale by counting the houses, of which there are only eight, and multiplying their number by some arbitrary figure for average inhabitants per house. A reasonably accurate figure could indeed be obtained by no more than the study of a map of suitable scale. The official census figures cannot be used because they cover a greater area. Where there is a town or village, counting is difficult but the census figures generally can be used. These vary according to the time of year, for the return includes all persons under any given roof on one particular day. A table compiled in this way was submitted to local authorities in the three counties and those in Lancashire and Westmorland were kind enough to reply and to supply modified figures. Coniston, Esthwaite, Windermere and Ullswater are the lakes in table 10 to which these corrections apply.

Mr H. Whittaker, Engineer and Surveyor to the North Lonsdale Rural District Council, writes: 'An equivalent of 384 houses are connected to the public sewer draining to Coniston Water; this includes hotels, schools, etc., and allowance for visitors. The average number of people per house is taken as 3·7 in this area.' He gives similar figures for Hawkshead and Near Sawrey which drain into Esthwaite and for Far Sawrey which drains into Windermere South Basin. Obviously these figures approach more nearly the information which limnologists seek than those based on total population. Actually the difference between the two is not great; for example, the above give a figure of 1420 for Coniston, the census one of 1100.

Mr G. A. Wade, Clerk to the Lakes Urban District Council, sent corrected census figures for the drainage area of Windermere.

Dewdney, Taylor and Wardhaugh (1959) found, in a survey of the Langdale Valley and its people, that 584 inhabitants (1951 census) could accommodate about 400 visitors. Probably a ratio of this order holds for all the rural areas, but the proportion of visitors is likely to be higher in places such as Windermere, Ambleside, Grasmere and Keswick where there are some large hotels.

The number of campers is generally less easy to judge, though there is

Table 10

The human population in the drainage areas of various lakes

Lake	Area in km²	Population in drainage area	People per km² lake surface
Ennerdale	2·91	32	11
Wastwater	2·91	50	17
Buttermere	0·94	32	34
Crummock	2·52	200	79
Derwentwater	5·35	724	135
Loweswater	0·64	100	156
Coniston	4·91	1,420	290
Ullswater	8·94	782	87
Bassenthwaite	5·35	5,912	1107
Windermere	14·72	13,292	896
N. basin only	8·16	5,314	651
S. basin only	6·66	7,978	1197
Esthwaite	1·0	1,239	1239

some definite information; for example, some 20,000 camper-nights are spent in Langdale each summer. This figure was kindly supplied by Mr C. H. D. Acland, Area Agent of the National Trust, which now owns most of this valley and permits camping only on a site which it provides.

It is impossible to state by how much the population in houses is increased by campers, but it is certain that conditions suitable for one are suitable for the other, and the ratio between the two may well be similar from one valley to another. It is not in the valleys with fewest houses that tents are to be found in noticeable profusion.

Nor can there be any exact relation between number of inhabitants and degree of enrichment of the lake, though it is likely to be proportionately higher with larger numbers because the big places will provide modern sewage for a greater percentage of the inhabitants. Of those relying on septic tanks, some may be enriching the lakes but little if the tanks are well sited with a good soak-away. On the other hand where there is rock or where the load is excessive there may be a copious supply of nutrients to the lake.

Even though, for these reasons, comparison between lakes based on the data in table 10 must be approximate it is nonetheless striking because the differences are so large.

CHAPTER 4

Physics

TEMPERATURE

Biologists are interested in temperature for two reasons: first it affects the distribution of animals; secondly, during the summer in temperate regions a lake becomes divided into separate layers, the upper warm (epilimnion), the lower cold (hypolimnion), and this influences the whole economy of the basin.

Macan (1963) devotes a chapter to the influence of temperature on distribution. Each species has an optimum temperature, though this is difficult to define exactly. Its range, or its numbers, may be limited because temperature:

1. reaches a high or low value that is lethal;
2. does not reach, or remain long enough at, a value high enough for successful breeding;
3. does not remain long enough above or below a value at which the population can compete successfully with a similar species with a different optimum, or maintain numbers sufficient to make good losses due to enemies.

Temperature may kill if it exceeds an upper or lower lethal level for no more than an hour or two, but within a narrow range of latitude the importance of this effect may be slight. In the British Isles low lethal temperature may affect the fauna in running water because streams commonly emerge from the ground at a temperature above that of the air in winter and retain a higher temperature for some distance. Nielsen (1950) states that certain Danish streams of this type harbour southern European species which cannot tolerate temperatures near freezing point. Such species are not able to find a comparable refuge in standing water except in a very small number that are warmed artificially (Feliksiak 1939). Every other piece of water falls to a temperature a degree or two above freezing point but thereafter, owing to the peculiar properties of water, goes no lower. The difference between a cold and a warm winter, or between a warm and a cold part of the country, is reflected in the thickness and the duration of ice, but not in the actual temperature of the water. Therefore in still water within the British Isles low lethal temperature is not an important ecological factor.

A high lethal temperature is also more important in running than in standing water. Streams are generally cool at the top and here are found various species that cannot tolerate the temperatures reached lower down where water and air temperature are in equilibrium. The most studied of these animals is *Crenobia* (*Planaria*) *alpina*, a recent and extensively documented work on which is that of Pattee (1965). A high lethal temperature is less important in lakes, first because the range of the maximum reached within an area such as the British Isles is less, and secondly because there is a cold hypolimnion offering a retreat when the epilimnion becomes too warm, though of course this retreat ceases to be available if oxygen disappears from the hypolimnion.

In general then, when temperature is the factor upon which the occurrence of a species in a lake depends, it does not act by reaching an intolerable level but in one of the other two ways mentioned. In both, duration is important as well as the actual temperature. One object of this preamble is, therefore, to stress the value of a continuous record compared with single readings. Mortimer (1952), concerned with temperature in relation to stratification and circulation, records that Wedderburn emphasized this point as long ago as 1911. One of Mortimer's figures (reproduced here as figure 23), illustrates this well. At 0900 hours on 11 June 1947 the surface temperature of Windermere was 14°C. It was in Mortimer's words: 'an exceptionally hot, calm day'. At 1600 hours the temperature was 16·7°C. It then fell under the influence of the wind, which mixed the thin warm superficial layer with the rest of the epilimnion, and at 0900 hours next day it was 14·4°C. Single readings taken at 0900 hours, the usual time, would have recorded a temperature 3·3°C below the maximum reached. Single readings of the surface temperature of a lake are of limited significance during the warming-up period.

At the risk of labouring the point, various records from Windermere that illustrate it are summarized in table 11. The surface temperature of Windermere was measured at the Wray Castle boathouse at about 0900 hours every morning from 1933 to 1950 and has been taken at the Ferry landing since then. From 1947 onwards a weekly reading at the surface, and at lower depths, has been taken at a buoy moored in the deepest part of the North Basin. Incidentally the first six years of these readings are quoted by Jenkin (1942), who also repeats some readings taken from West for the year 1908. These are so low that only a phenomenal year could have produced them. Dr J. W. G. Lund has informed me the records kept locally by a private individual do not show 1908 to have been exceptional, and it may safely be concluded that these readings are unreliable.

While the lake is warming up, the temperature taken from the edge is nearly always higher than at the surface in the middle, sometimes by more than 2°C. In winter it may be lower: the sampling point has sometimes been covered with ice when the middle of the lake is open. The temperature is generally higher at the surface than at 5 m below it (table 11). It is evident that the temperature at 5 m gives the best indication of the con-

Table 11

Maximum temperature of Windermere 1947–1964

Daily spot readings at about 0900

The maximum temperature has been divided into intervals of 1°C, and the figures show the number of years when the maximum fell into a given interval. For example there were six years when the maximum temperature at the edge exceeded 20°C but did not reach 21°C

	16	17	18	19	20	21	22	23°C
Boathouse 1947–1950								
Ferry Landing 1950–1964	1	1	2	3	6	1	2	2
Surface at N. Basin buoy	3	3	2	3	5	0	1	1
5 m at N Basin buoy	5	2	5	3	3	0	0	0

ditions in which organisms in the epilimnion, or on exposed shores bathed by epilimnion water, are living. It is the figure used in table 12 for which seven years, each unusual in some way, have been selected. Three extremes of high and one of low temperature which occurred in other years are shown in the last line but two.

More frequent observations were obtained by Mortimer from 1947 onwards. In 1947 (Mortimer 1952) he used a thermo-electric thermometer by means of which he could obtain readings at eight stations within the space of two hours. Later he slung a series of thermistors at different depths from a buoy and connected each one to a recorder in the laboratory. In this way, he obtained a record that, to all intents and purposes, was continuous.

Mortimer's object in measuring temperature was to solve a problem that he had uncovered during the course of an earlier study (Mortimer 1941–2). Investigating the substances that diffuse out of the mud when the hypolimnion becomes de-oxygenated, he had found that these become distributed throughout that region of the lake much faster than could be accounted for by molecular diffusion, a slow process. Vertical eddies were obviously stirring the hypolimnion water, and there was reason to believe that these were set up by horizontal currents. It had long been known that, when a wind blows over a lake, it tilts the surface of the water only slightly, so slightly that measurement requires a refined technique. In contrast the thermocline is tilted appreciably; the epilimnion is blown down towards the leeward end of the lake and the increase in its depth may be measurable in metres. The surface of the hypolimnion slopes upwards towards the windward end of the lake. When the wind ceases, the hypolimnion has been raised above the equilibrium level at the windward end and depressed below it at the other end of the basin. The whole of the

Table 12

Temperature of Windermere: monthly maximum and minimum at depth of 5 m at North Basin buoy. Years selected from the period 1947–1964

Year	Peculiarity		Jan.	Feb.	Mar.	Apr.	May	June	July	Aug.	Sept.	Oct.	Nov.	Dec.
1947	spring cold	max.	5·6	4·0	2·6	5·9	10·7	16·9	17·4	20·9	20·9	14·1	10·0	7·0°C
	late summer warm	min.	4·7	2·1	2·4	3·0	5·9	11·7	15·1	18·9	14·2	11·8	7·6	6·3
1949	warm	max.	6·0	5·6	6·3	8·4	13·2	16·9	19·0	18·0	18·0	15·8	9·6	7·5
		min.	5·7	5·3	5·1	6·3	9·8	13·2	17·2	16·2	16·2	11·3	7·7	6·6
1954	wet (fig. 11)	max.	6·4	4·8	5·5	8·0	12·0	14·0	15·1	16·0	14·3	11·8	10·7	7·7
		min.	5·6	3·9	3·6	5·6	7·7	13·0	13·5	13·4	12·1	11·3	7·8	6·8
1957	warm spring and	max.	6·3	5·8	7·0	9·5	12·0	18·0	19·0	18·1	14·5	11·6	9·8	6·9
	early summer	min.	5·7	5·3	5·1	7·0	10·0	14·0	17·1	12·2	11·3	10·3	6·6	4·8
1959	warm summer	max.	6·1	3·6	5·2	7·5	13·0	17·1	18·0	19·0	19·0	15·4	10·6	8·4
	and autumn	min.	3·9	3·2	4·5	5·7	8·4	14·3	17·3	17·5	15·5	12·4	8·6	6·5
1962	cold	max.	5·0	4·6	3·9	7·2	10·0	14·0	16·0	15·3	14·0	12·5	9·8	7·9
		min.	4·0	3·7	3·1	4·3	8·0	11·0	13·1	13·7	12·7	10·6	7·5	4·8
1963	very cold	max.	3·8	2·5	3·7	8·0	9·0	14·5	16·0	16·0	14·4	11·8	10·4	7·7
	winter	min.	1·7	1·9	2·4	4·2	7·5	12·0	13·4	14·6	12·6	10·6	8·2	5·6
Extreme (if not included above)			6·5	–	–	–	13·3	–	20·0	–	–	8·6	–	–
			(1955)				(1948)		(1951)			(1952)		
Number of times highest temp. of year recorded								1	10	6	1			
Number of times highest minimum temp. of year recorded									8	10				

46

hypolimnion then reverts to the equilibrium position but, in doing so, gains momentum which carries it beyond the equilibrium point so that the thermocline becomes tilted in the opposite sense. It then swings back and

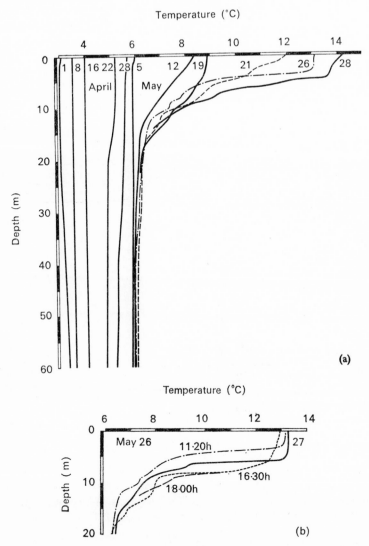

18. Vertical distribution of temperature at station B, Windermere northern basin (*a*) at approximately weekly intervals from 1 April to 19 May 1947, and thereafter more frequently until 28 May, (*b*) on three occasions on 26 May (Mortimer, C. H. (1952), *Phil. Trans. R. Soc.* **236**).

again passes the equilibrium position, and this see-saw swing continues with steadily decreasing amplitude, probably for days, though the wind has generally sprung up again to impose a new pattern before the earlier oscillation has completely died down. As the epilimnion and hypolimnion roll to and fro, the water streams from one end of the lake to the other and then back again. What was not known, and what Mortimer wished to find out, was the rate of this flow, particularly in the hypolimnion. Wind sets up a surface current in the epilimnion and a current in the opposite direction lower down brings back the water carried by it to one end of the lake. It has been suggested that the lower current will drag along surface hypolimnion water owing to friction and set up circulation in the hypolimnion, but there are objections to this idea. Mortimer believed the current to be so slow that measurement by means of any existing flow meter

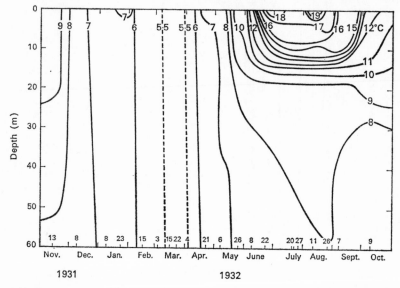

19. Temperature at different depths in the north basin of Windermere throughout the year from November 1931 to October 1932 (Jenkin, P. M. (1942), *J. Anim. Ecol.* **11**).

would be difficult, and assumed that an indirect approach by measuring temperature changes would reveal its rate more readily.

Mortimer's (1952) figures also give a clear picture of the process of warming and stratification in a lake. The early part of 1947 was unusually cold (table 12), and on 1 April the top 25 metres of the lake were at 3°C. Lower down there was slight warming, an example of inverse winter stratification (fig. 18). The lake warmed up during the next five weeks, with, however, little difference between the top and the bottom and by 5 May it was more or less uniform at 6°C. Thereafter the surface layers warmed up rapidly to form an epilimnion, which by 28 May was about 5 m deep and about 14°C in temperature, there being a small decrease from the surface downwards. Temperature decreased rapidly down to about 10 m where it was about 8·5°C and then more slowly to about 16 m below which it was uniformly just above 6°C. The inset in fig. 18 is relevant to Mortimer's main problem and may be ignored for the moment.

Mortimer's records do not extend beyond June and for a picture of a whole season we must refer to Jenkin's earlier readings, obtained with the old-fashioned reversing thermometer (fig. 19). The winter of 1931–2 was milder than that of 1946–7. The lake was losing heat until mid-March, and stratification was not well established till the end of May. During June there was a rapid warming of the epilimnion, in July conditions were stable, in August there was a small gain in heat and in September cooling started.

An overturn was studied by Mortimer, but before that is described, his main findings will be examined. Between 11.20 and 16.30 hours on 26 May, that part of the graph on the inset of fig. 18 which is most nearly horizontal, in other words the thermocline, had sunk about 4 m. Sufficient heat to cause this could not have been absorbed in the time, and it could only be due to a tilt. On 28 May measurements were made at the four stations in the North Basin (P, B, R, S on fig. 15) after a force 6 wind had been blowing for about four hours from SSE. The tilt of the thermocline was 1·2 m/km (fig. 20 upper). Next morning, after a calm night, the epilimnion had rolled back and the thermocline was tilted in the opposite sense (fig. 20 lower). This was the beginning of a calm sunny spell which lasted until 3 June, and observations at station B were continued throughout this period and up to 21 June (fig. 21). The lines of vertical dots on this figure show when the measurements were made. The thermocline continued to oscillate after the wind had dropped, and the epilimnion became warmer. On 3 June the surface temperature had reached 21°C, and at lower levels it decreased at a fairly regular rate of some 1·6°C/m to 11°C at 6·5 m. There was no clearly defined thermocline. The wind then sprang up from the north-west and reached force 7–8 on 9 June. As a result the temperature of the epilimnion became uniform at 14·7°C, and there was a sharp thermocline with a drop of 4°C between 6 and 8 m. This change was due almost entirely to mixing, and the mean temperature of the top 19 m had increased from 13·5°C on 3 June only to 13·7°C on 7 June. On the latter date there was a distinct epilimnion, a sharp thermocline, and below it a zone in which the temperature dropped slowly down to a depth of about 20 m, beyond which the temperature changed only slightly down to the bottom. This lowest uniform zone is the hypolimnion, and the zone above it, including the thermocline, is treated by Mortimer as the metalimnion.

A transverse series of readings running through station B showed that a cross wind produced a lateral tilting of the thermocline, which prompted Mortimer to remark that 'the structure of the water column at B is subject to influences of considerable complexity'.

During the gale of 9 June, and later when it was moderating, temperatures were recorded at the four stations (fig. 22). The epilimnion was

E   49

mixed, as already noted, and blown down to the leeward end of the lake into a wedge which left metalimnion water exposed at the surface at the windward end. Some of this cold water was mixed with the epilimnion,

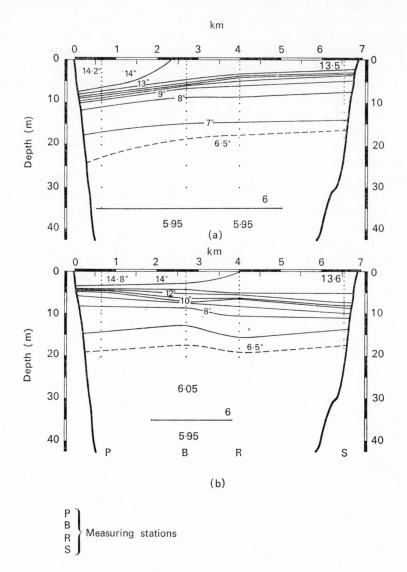

Note : The vertical scale is 100x the horizontal

20. Distribution of temperature in a longitudinal section of Windermere, northern basin: (*a*) on 28 May 1947 (15.10 to 17.07 h.) after an east-south-east wind had been blowing for about four hours, (*b*) on 29 May 1947 (09.07 to 10.58 h.) after a calm night. The positions of the measuring stations are shown on fig. 15 (Mortimer, C. H. (1952), *Phil. Trans. R. Soc.* **236**).

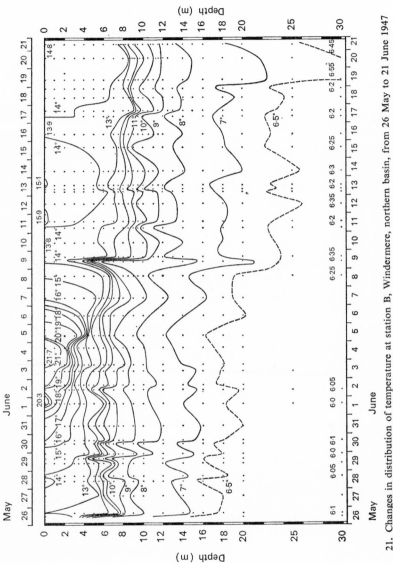

21. Changes in distribution of temperature at station B, Windermere, northern basin, from 26 May to 21 June 1947 (Mortimer, C. H. (1952), *Phil. Trans. R. Soc.* **236**).

51

which in consequence was colder at the north end when it once more covered the whole basin. In this way strong winds deepen the epilimnion and cause a reduction in its average temperature.

Hourly observations were made at station P during the calm days which followed the gale (fig. 23). The thermocline oscillated with an amplitude of 2 to 3 m and a period of 18–19 hours, and there was a smaller oscillation with a period of 2 to 3 hours. Considerable oscillations in lower layers were not in phase with those at thermocline level.

The next stage of the work was in a field of physics and mathematics into which it is not proposed to follow Mortimer here. Calculations based first on the assumption that the lake consisted of two layers gave results not in accord with the observed facts. A better fit was obtained when three layers, epilimnion, metalimnion and hypolimnion were postulated. The currents set up in these three layers by the gale of 9 June and by the subsequent seiche are shown in fig. 24. In the epilimnion a speed of about 6 cm/s was attained and in the hypolimnion the current attained about half that speed.

The overturn was studied in 1951 (Mortimer 1955). On 5 November the temperature of the epilimnion ranged from 9·6 to 9·79°C, that of the hypolimnion from 6·6 to 6·76°C and there was a thermocline at about 30 m. Wind depressed the thermocline but, with this slight difference in density, only a slight seiche was set up. It was noted that the resulting flow was deflected to the right and this was attributed to the earth's rotation (Coriolis force). A gale started on 7 December and by the following day temperature was uniform to within 0.08°C from top to bottom, the mean being 7·44°C. A few days later the temperature in the rivers Rothay and Brathay was only just above 4°C and this cold water was accumulating at the bottom of the lake.

To lay before the reader, who has been studying Mortimer's portrayal of the ever-changing temperature pattern, profiles based on one set of readings for each lake (figs. 30 and 31) is like inviting a keen cinema-goer to an evening's entertainment with the magic lantern. However, they are all that is available, and I am most grateful to Dr Vera Collins, who made the observations and has put them at my disposal before her own publication has appeared.

Stratification is slight in Bassenthwaite and Derwentwater, both lakes which are broader and more shallow than the others (table 4). Strong wind might leave either of them completely mixed but how strong a wind would be required for this and how often complete mixing occurs during a summer cannot be stated. Whether a sunny still period in May or June ever produces stratification stable enough to remain for the rest of the summer, and if so, how often this happens are additional questions to which only further work will furnish the answers.

The differences between the other lakes are no greater than could have

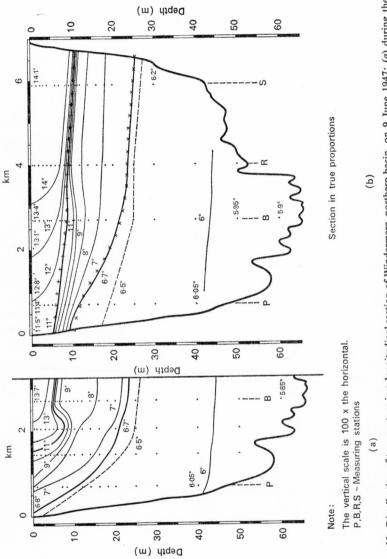

Section in true proportions

(b)

Note:

The vertical scale is 100 x the horizontal.
P,B,R,S – Measuring stations

(a)

22. Distribution of temperature in a longitudinal section of Windermere, northern basin, on 9 June 1947: (*a*) during the morning (10.30 to 12.10 h.) at the height of a north-westerly gale force 7 to 8; (*b*) during the afternoon (15.25 to 17.14 h.) as the gale was moderating (Mortimer, C. H. (1952), *Phil. Trans. R. Soc.* **236**).

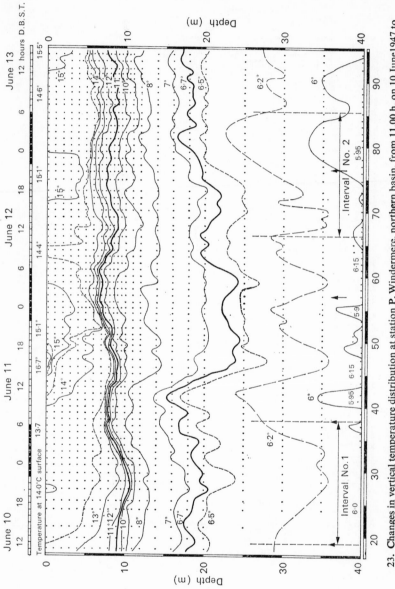

23. Changes in vertical temperature distribution at station **P**, Windermere, northern basin, from 11.00 h. on 10 June 1947 to 15.00 h. on 13 June 1947 (Mortimer, C. H. (1952), *Phil. Trans. R. Soc.* **236**).

been produced by different weather conditions immediately preceding sampling. The sharp thermocline with uniform conditions above and below, as is seen in Esthwaite for example, is probably the result of a strong wind (cf. fig. 21). The more gradual drop in Wastwater was probably produced by fine weather and moderate wind. Records of wind speed are not available for the other lakes, but it is a matter of common observation by members of the F.B.A. staff that Ennerdale and Wastwater, the lakes which lie close to the highest mountains and whose valleys open towards the sea, are often too rough for work on days when Windermere is calm. They are probably under the influence of strong winds much more continuously than Windermere, but here too a comparison of what the result of this may be on the temperature profile requires more frequent measurements.

It is instructive to compare the Lake District lakes, which, lying on the Atlantic side of the continent, come frequently under the influence of depressions, with the Carinthian lakes, which are situated in a region of the Alps where strong winds are much less frequent. They were studied by Dr I. Findenegg, and the details here are taken from an account which he wrote in 1953 for the congress of the International Association of Limnology, though some of the results had been published 20 years earlier. The Carinthian lakes are of about the same size as those in the English Lake District; the largest, the Wörthersee, is of about the same length as Windermere, though a little broader and deeper. All these Austrian lakes are higher than the English ones, the lowest, the Wörthersee again, being 439 m above sea level (cf. table 4), but on the other hand they lie some 7° further south. In spring, by the time the ice covering the lakes has melted, the sun is high enough to start warming the epilimnion at once. At this time of year there are generally breezes caused by the warming of the land and these are sufficient to mix the warmest water at the surface with the colder layers immediately below, but not to mix the whole lake as Windermere was mixed in the years illustrated in figs. 18 and 19. Findenegg stresses the importance of some wind, because the sun warms no more than the top metre and, if this were left undisturbed, much of the heat would be lost by radiation at night. A sharp thermocline is soon established, and the difference in density between epilimnion and hypolimnion is such that the occasional summer storm will not mix them. The absence of strong winds also has the effect of producing a comparatively shallow epilimnion, which, combined with the long hours of sunshine characteristic of the region, leads to an unusual accumulation of heat. Temperatures of 22–24°C, and up to 28°C in bays, are usual and the lakes are popular for bathing.

The autumn is a particularly windless time and, as the surface of the lake cools, the water there becomes denser than it is lower down and the convection caused by its sinking brings about mixing. If the previous winter's temperature has been 4°C, surface water cooling to this temperature may not mix with the hypolimnion water, and stirring by winter winds does not take place because ice may form soon after the surface has reached 4°C. Once this phenomenon has occurred, mixing in subsequent years becomes less likely because a difference in density due to substances in solution hinders it. The epilimnion is depleted of dissolved substances by organisms which, on their death, carry them down to the hypolimnion where they are released by decomposition.

The hypolimnion of some Carinthian lakes is permanently isolated from the surface waters. Findenegg has called them meromictic lakes. This condition depends to some extent on a small surface area in relation to depth, on a small inflow, and on a good supply of substances carried down to the hypolimnion from a productive epilimnion. All these features are exhibited by the Carinthian lakes but the main condition is the stillness of the region and the rarity of strong winds.

The lower layer of a meromictic lake, the monimolimnion, is devoid of oxygen. The temperature in five of the lakes ranges from 4·2–5·1°C. This is attributed to occasional storms which, though not strong enough to mix the monimolimnion—with its comparatively high density due to solutes —completely with the rest of the lake, do effect some mixing with warmer water from above.

Since their discovery in Austria, meromictic lakes have been recorded in various parts of the world, but it is safe to assume that the condition will not occur in the English Lake District where storms are the rule and ice-cover the exception.

LIGHT

The transparency of the water was one of the criteria used originally by Pearsall (1921) to assign each lake to a place in the series (table 13) (fig. 25).

Table 13

Depth at which a Secchi disc 7 cm in diameter can just be seen (Pearsall 1921)

	m
Wastwater	9
Ennerdale	8·3
Buttermere	8·0
Crummock	8·0
Haweswater	5·8
Derwentwater	5·5
Bassenthwaite	2·2
Coniston	5·4
Windermere	5·5
Ullswater	5·4
Esthwaite	3·1

Modern technique has not lead to an alteration in this arrangement but it has advanced considerably since Pearsall's time, and fig. 25 in which the penetration of light into two productive and one unproductive lake is contrasted, is based on data kindly supplied by Dr J. F. Talling. The order agrees well with that based on other factors (table 3) except that Bassenthwaite comes out of place. Its low transparency is due to dissolved and suspended humic materials derived from the bogs in which its main inflow originates (see p. 16). The low figure for Esthwaite is partly due to the same cause, but otherwise Pearsall assumed that the light was absorbed by plankton and that the distance to which a given proportion of the surface intensity penetrated was a measure of the amount of plankton. That this was a valid assumption was shown some years later by the data presented here in fig. 26 (Pearsall and Ullyott 1934).

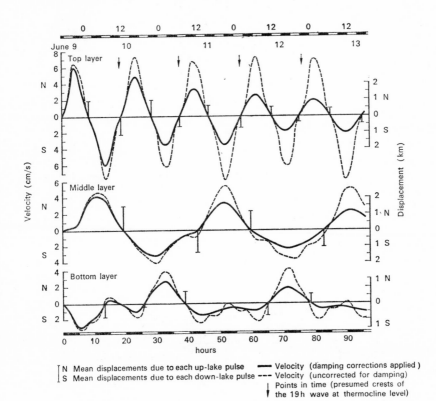

N Mean displacements due to each up-lake pulse	── Velocity (damping corrections applied)
S Mean displacements due to each down-lake pulse	--- Velocity (uncorrected for damping)
	Points in time (presumed crests of the 19 h wave at thermocline level)

24. Theoretical horizontal components of velocity and displacement in the top, middle and bottom layers at the uninode of Windermere, northern basin, during the period 9–13 June 1947 (Mortimer, C. H. (1952), *Phil. Trans. R. Soc.* **236**).

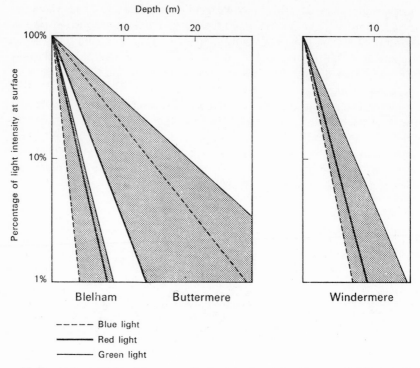

25. Penetration of light of three wave-lengths into three lakes (data supplied by Dr J. F. Talling).

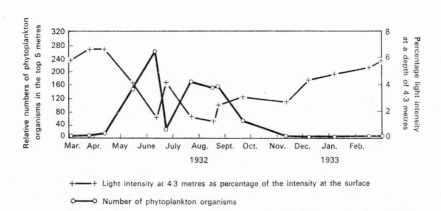

26. Light intensity at the limit of aquatic vegetation (4·3 metres) compared with the numbers of phytoplankton organisms in the top 5 metres (Windermere, March 1932–March 1933) (Pearsall, W. H. and Ullyott, P. (1934), *J. exp. Biol.* **11**).

QH
138
L35
M 12 │ 16,425

CAMROSE LUTHERAN COLLEGE
LIBRARY

CHAPTER 5

Chemistry

INORGANIC SUBSTANCES

The identity and the amounts of solutes entering a lake are the domain of the analyst. Within a lake the concentrations of some of them are influenced by the activities of organisms, notably algae. There has always been a close collaboration between the chemists and the biologists on the staff of the Freshwater Biological Association, and the results of many analyses are to be found in biological papers. No attempt is made here to divorce the two subjects, and some information on chemistry is included in later chapters.

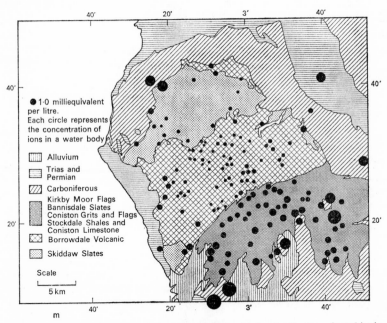

27. Outlines of the geology of the Lake District, and the distribution of total ionic concentration in the surface waters (Mackereth, F. J. H. (1957), *Proc. Linn. Soc. Lond.* **167**).

Sodium, calcium, magnesium, potassium, hydrogen, bicarbonate, chloride, sulphate and nitrate contribute 99 per cent of the ions dissolved in the surface waters of the Lake District (Mackereth 1957). Fig. 27 shows the total concentration of ions, indicated by the size of the circle, in lakes and smaller bodies of water within the Lake District, defined in geological terms, and just outside it. Concentration is low (0·15 to 0·2 milli-equivalents/litre) on the Ordovician rocks, both Skiddaw Slates and Borrowdale Volcanics. It is higher (0·5 to 2·0 m–e/l) on the Silurian rocks, and higher still (2·0 to 5·0 m–e/l) on the younger formations which encircle the Lake District. However, waters lying to the west near the sea have a higher concentration than waters of the same type further inland.

The main ions on the Ordovician rocks are sodium, hydrogen, magnesium, chloride and sulphate. On the Silurian rocks calcium and bicarbonate rank equal with these, and on the younger rocks they supersede them. Except in the extreme west, where sodium and chloride are the most abundant ions, there is an increase in the proportion of calcium and bicarbonate as total ionic content increases. These substances are present in rain, or are dissolved out of the soil by rainwater.

Gorham (1955, 1958) studied rain and found in it all the ions mentioned in concentrations varying widely with season, and with the force and direction of wind and with the duration of rainfall. He plotted the concentration of each ion against that of every other and found that sodium and chloride always maintained a ratio which suggested that their origin was from the sea. The higher concentrations of these two ions in waters near the coast, and in rain brought by south-westerly winds provide strong corroboration for this interpretation. Magnesium also comes from sea spray. On the other hand, there is a rough inverse relation between sulphate and chloride, and when the concentration of one is high, that of the other never is. The highest concentration of sulphate is in rain falling when the wind blows from the south-east, which, together with its correlation with rain containing soot, indicates that it is derived from pollution of the atmosphere by the industrial area of south Lancashire. Calcium and potassium both show some correlation with sulphate.

Gorham calculated that during the course of twelve months 92 Kg/ha chloride and 24 Kg/ha sulphate fell on the hills which lie to the west of Esthwaite.

The relative importance of ions from rain and from the soil was summarized by Gorham (1958) in a figure reproduced here as fig. 28. Throughout the Lake District, and on the more calcareous rocks beyond, nearly all the sodium is brought by the rain. On the Ordovician rocks more than 50 per cent of all the ions on fig. 28 may be derived from this source, but on other formations it is less important. Calcium and magnesium are of mainly terrestrial origin in all but the waters with the lowest total content of ions. The proportion of potassium supplied by the rain falls less steeply with increasing ionic concentration and that of sulphate less steeply still.

A curiosity, unique as far as is known, is the salt spring near the head of Derwentwater. It has been known for a long time and, in the eighteenth and early nineteenth century, was visited for the medicinal properties which it was believed to possess. Kipling (1961) has written a short account of its

Chemistry

a. Upland tarns on hard volcanic rocks

b. Upland and lowland tarns and lakes
   on Silurian slates and flags

c. Lowland tarns on softer sedimentary rocks
   of the Carboniferous and Triassic periods

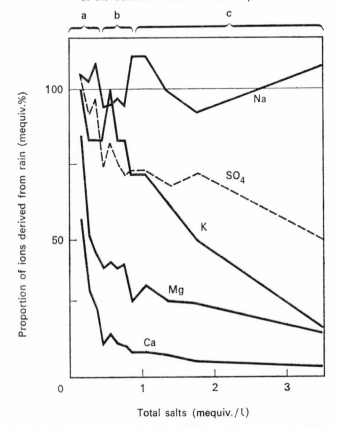

28. The proportions of ions derived from rain in tarn and lake waters
of different salt content on various geological substrata (Gorham, E.
(1958), *Phil. Trans. R. Soc.* **241**).

history and Mr F. J. H. Mackereth has very kindly supplied me with the
following analysis:

	p.p.m
Sodium	4,700
Calcium	4,400
Chloride	14,700
Potassium	68
Carbonate	47

The water is about half as saline as the sea but with a much greater amount

61

of calcium. The flow from the spring is slight and its influence on fauna and flora beyond its immediate vicinity is negligible.

Ideally, the next stage in an account of the chemistry of the waters of the Lake District would be a statement showing the amount washed in and the amount washed out of each lake in a year. This ideal, however, has so far proved unattainable in a district where the rainfall varies so irregularly. Heavy rain can cause marked changes in flow within a period of hours, and it is unlikely that concentration of any substance remains constant during this period. Only a continuous record of flow and concentration would give an accurate answer, and the apparatus for this has not yet been devised.

Notwithstanding the difficulties, Mortimer (1938) compiled a nitrogen balance sheet for the Lake District lakes. Once a month he analysed for nitrate nitrogen and organic nitrogen samples from the rivers Rothay and Brathay, Cunsey Beck and the outfall of the Windermere–Bowness Sewage works, all inflows into Windermere. From the average flow and this single analysis the amount of nitrogen entering the lake each month was calculated. The irregularity of the results confirms the initial suspicion that such figures can give no more than a rough approximation. The conclusion is that 326 metric tons of nitrogen enter Windermere each year and 318 leave it. However, Mortimer believes that his figure for input is too low, and postulates more nitrogen from the following sources:

1. Streams. Mortimer assumed that concentration in streams was the same as in the large inflows, but in some analyses which he made it was higher.
2. Sewage entering the lake by routes other than the sewage works whose outfalls enter one of the inflows analysed.
3. Dead leaves. These may play a larger part in the economy of a lake than has hitherto been suspected. There is little information about how many dead leaves fall to the bottom of a lake each year, and it is difficult to see how the information could be obtained. Nor is much known about the decomposition of dead leaves. Hynes (1961) notes that most of the production of animal tissue in a stream takes place in winter and he suggests that one possible explanation is that the main source of food is dead leaves, which are abundant only during that period. He believes that the animals cannot utilize dead leaves directly but derive nourishment from the bacteria and fungi that are breaking them up. The present writer recollects that the sorting of a collection from a stream was made more difficult in winter by packets of dead leaves, which appeared to be continually renewed. There was evidence that wind moved dead leaves from place to place throughout the winter and that a leaf ceased to circulate either when it fell into water or when grass began to grow and other plants began to put forth leaves in the following

spring. This casual unsystematic observation is put on record because it suggests the possibility that a high proportion of the leaves in a drainage area ultimately come to rest in a lake, if there be one.

Notwithstanding the probably approximate nature of Mortimer's results, a small number of analyses of the average nitrogen content in main inflows and outflows from nine other lakes gave figures sufficiently consistent to suggest a fair degree of accuracy. In five unproductive lakes, Wastwater, Ennerdale, Buttermere, Crummock and Haweswater, the average nitrogen content of the inflows ranged from 0·18 to 0·26 p.p.m., and that of the outflows was generally a trifle higher. Thirlmere had higher values, 0·51 p.p.m. in the inflow and 0·53 p.p.m. in the outflow. In three productive lakes, Loweswater, Esthwaite and Blelham, the nitrogen content of the inflows ranged from 0·62 to 1·16, and that of the outflow was a little over half that in the inflow. In other words, nitrogen was accumulating in these productive lakes as it was in Windermere, if Mortimer was right in assuming that his figure for input was too low.

The concentration of any substance in a lake will, subject to the proviso about to be examined, reflect the average composition of the water flowing in during the time of retention in the lake. If Windermere could be drained, it would take nine months' average rainfall to fill it again. Obviously when the lake is stratified, theoretical retention time of inflowing water, provided it is warmer than the hypolimnion, is much less. In practice it depends on relative temperature of inflowing and lake water and on wind strength and direction.

The proviso mentioned is that the substance is not taken out of circulation by organisms or by non-living substances. Nitrate, phosphate and silicate are, as will be described shortly, utilized by algae, and the uptake of phosphate by *Asterionella* has been studied in a series of careful culture experiments by Mackereth (1953). He was puzzled at the outset to reconcile his own observation that proliferation of *Asterionella* does not affect the concentration of phosphate in the lake with Gardiner's demonstration that it produced a proportional fall. One difference between the two sets of observations was that Mackereth's were made in Windermere where the concentration of phosphate ranged between less than $1·0$ μg P/l and $2·0$ μg P/l, Gardiner's in a water where the initial concentration was 66 μg P/l.

Mackereth found that the limiting concentration of phosphate in cultures was low, about $0·06$ μg/10^6 cells, and that growth brought to an end by depletion to this level was resumed when more was added. In Windermere phosphate is not a limiting factor, for, as will be seen later, silicate reaches a critical concentration before it does. At low external concentrations, such as that in Windermere, *Asterionella* takes up phosphate and stores it in amounts well in excess of requirements. When a cell divides the store is shared equally. In this way a population can increase rapidly until checked by deficiency of something else, without taking up further supplies of phosphates from the water. Mackereth suggests that, when phosphate in solution is plentiful, an increasing population maintains its store at or near the original level and replenishes the loss at each division from the water. In this way he explains the different results of Gardiner and himself.

Other findings not relevant to the present theme but to discussions in the next chapter are that *Asterionella* cannot utilize organic phosphorus. Also it cannot utilize phosphate in distilled water, in Windermere water that has been evaporated to one-fifth of the original volume, or in distilled water to which has been added the proportion of salts present in Windermere. Phosphate is removed from a culture of water from Wastwater. The rate of uptake declines with increase or decrease of pH from the neutral point.

The normal uptake of phosphate from lake water evidently depends on the presence of some substance whose nature has not been discovered.

Phosphate is also removed from solution by organic compounds, which are noticed again later in this chapter, and by a ferric complex in the mud. This was the subject of a long investigation by Mortimer (1941–2) into events in Esthwaite and other lakes and in an aquarium. In 1939 Esthwaite became stratified during the month of May, and by mid-July oxygen had disappeared from the lower layers of the hypolimnion. It was absorbed by the mud, quickly at first and then more slowly as the gradient at the mud surface fell. At a concentration of about 0.5 mg/l O_2 ferric was replaced by ferrous iron at the mud surface and diffusion of this reduced iron into the water removed the last traces of oxygen there. At the same time the total concentration of dissolved salts increased by about 60 mg/l. The concentration of phosphate and of ammonia increased a hundredfold, and only one-tenth of this amount could have been produced by the reduction of nitrate. Silicate rose from a concentration of 1 to 3 mg/l. These substances are apparently held by a ferric complex and released when deoxygenation leads to the appearance of the soluble ferrous ion.

Mortimer observed that, after they had been released at the mud surface, these substances spread through the hypolimnion much faster than could be accounted for by diffusion, and his successful search for an explanation of this was described in the previous chapter.

When the lake turned over early in October, the iron was oxidized to the ferric state, the complex reformed and the substances that had been liberated were adsorbed once more.

A small number of analyses of other lakes showed in Blelham, Rydal and Loweswater a cycle of events similar to that in Esthwaite, but in Windermere, Crummock and Ennerdale there was no deoxygenation and no increase in the concentration of ions in the hypolimnion.

Mortimer measured the depth of the brown oxidized layer of mud in winter and found it thickest in Ennerdale and progressively less thick in

Crummock, Windermere South Basin, Windermere North Basin and Esthwaite. This corresponds to the order of production except that Windermere South Basin is more productive than the North. Mortimer attributed this to the greater abundance of the dead leaves, already mentioned, in the North Basin, an important point to which further reference is made in the next section.

Mortimer writes that the deoxygenation of the hypolimnion in the second but not in the first 'suggests that the conditions found in Windermere and Esthwaite may be considered as representative of two fundamentally different lake types'. This is the continental view, with which Mortimer, who had taken a doctorate of philosophy in Germany, was thoroughly familiar. It cannot be denied that to an animal living in the mud, the presence or absence of oxygen during the summer is of importance. However, a study of plankton, fish and the benthic organisms above the thermocline does not indicate a fundamental difference between Esthwaite and Windermere and Pearsall's idea of a series is more valuable.

Heron's (1961) discussion of the fluctuations in the concentration of phosphate, silicate and nitrate is based on weekly samples, and centres about four similar diagrams, one each for Windermere North Basin, Windermere South Basin, Esthwaite and Blelham. The first of these is reproduced as fig. 29. All three substances reach their highest concentration in winter and their lowest in summer, a fluctuation obviously associated with the number of algae. Small rises in summer are associated with rain. The decrease of phosphate starts before that of nitrate, which is attributed to the abstraction and storage described by Mackereth. The rise in winter is irregular both in time and in amplitude. The height reached in both basins of Windermere was not greatly different in the two winters, but in Esthwaite and Blelham it was higher in the second than in the first, probably because during the second there were fewer diatoms and more rain. Heron points out that the absence of a regular autumn rise in Esthwaite indicates that the release of phosphate in the hypolimnion when oxygen disappears is of slight biological significance. The fluctuation of silicate is more regular than that of phosphate and nitrate. Rises in summer are due not only to floods but also to the death of diatoms.

The other substance which fluctuates markedly is oxygen, particularly in the hypolimnion of stratified lakes. Fig. 30 shows that in Esthwaite, Loweswater, Rydal, Grasmere and Blelham, all or nearly all the oxygen is removed, as had been shown earlier by Mortimer, and in Windermere South Basin and in the two broad shallow lakes, Derwentwater and Bassenthwaite, there is some lowering of the oxygen concentration in the hypolimnion. In the rest there is very little (fig. 31).

The substances whose concentrations do not vary greatly with the changing season, or the wax and wane of algal populations, are the common ones. Presumably they are present in quantity far in excess of the

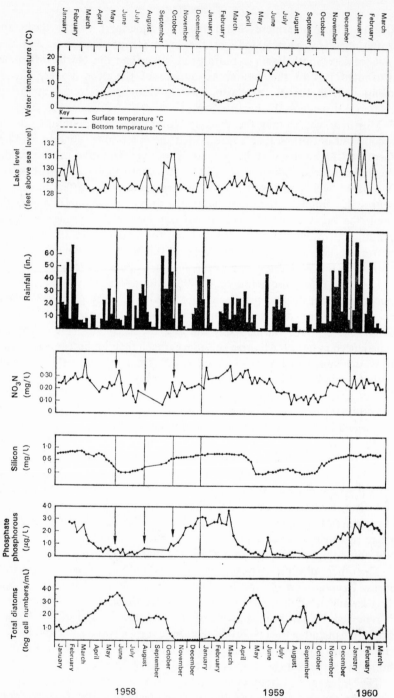

29. Windermere, north basin. Variation with time of water temperature, lake level, rainfall, nitrate nitrogen, silicon, phosphate phosphorus and total diatom population (Heron, J. (1961) *Limnol. Oceanogr.* 6).

requirements of the various algae. Lund (1957) states that their concentration is unlikely to rise to twice or sink to half the average, and the figures in the record book at Ferry House (averaged here in table 14), bear this out. The maximum and minimum are generally much closer to the average than this. Nitrate is an exception, as is to be expected, for it is an important

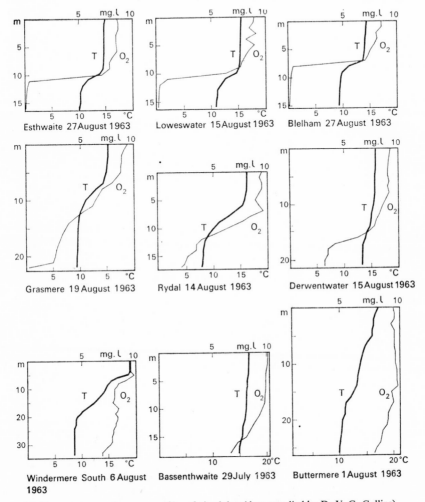

30. Temperature and oxygen profiles of nine lakes (data supplied by Dr V. G. Collins).

plant nutrient present in smaller amount than anything else in table 14. There are three other exceptions to Lund's statement but the margin is very small, and it may be left to stand as a rough general principle.

When the lakes are compared, it is seen that the concentration of calcium bicarbonate increases steadily along the series from unproductive to productive lakes, but the difference between the lowest and the highest

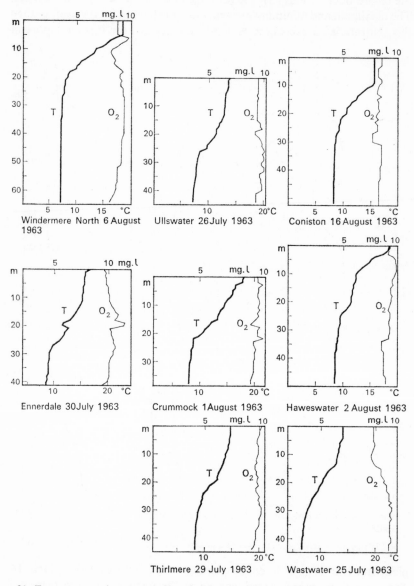

31. Temperature and oxygen profiles of eight lakes (data supplied by Dr V. G. Collins).

68

Table 14

Average concentration in mg/l of the major ions in the English lakes and in a hard-water lake. The averages for most of the English lakes are based on 5 or 6 readings taken at different times of year

	Ca	Mg	Na	K	HCO$_3$	Cl	SO$_4$	NO$_3$
Esrom Lake	42	5·6	12	No data	140	22	8·2	0
Esthwaite	8·3	3·5	4·7	0·90	18·3	7·6	9·9	0·78
Windermere S.	6·2	0·70	3·8	0·59	11·0	6·7	7·6	1·2
Windermere N.	5·7	0·61	3·5	0·51	9·7	6·6	6·9	1·2
Coniston	6·1	0·89	4·4	0·66	10·8	7·8	8·0	1·1
Ullswater	5·7	0·89	3·3	0·35	12·7	5·5	6·8	0·75
Bassenthwaite	5·3	1·2	5·0	0·66	10·0	9·1	7·4	1·1
Derwentwater	4·5	0·46	4·8	0·39	5·4	10·1	4·8	0·44
Crummock	2·1	0·78	3·7	0·31	2·9	6·8	4·5	0·35
Buttermere	2·1	0·72	3·5	0·27	2·6	6·9	4·1	0·48
Ennerdale	2·2	0·79	3·8	0·39	3·5	6·7	4·5	0·62
Wastwater	2·4	0·68	3·6	0·35	3·2	5·9	4·8	0·62
Thirlmere	3·3	0·67	3·1	0·31	4·1	5·4	6·0	0·62

is not great, and much less than that between the highest and the highly productive Esrom Lake. Of the rest only sulphate shows a tendency to be more concentrated in the productive than in the unproductive lakes, though a high value is recorded from Thirlmere. The amounts of magnesium, sodium and chloride in Windermere and Ennerdale are identical or very close, though all are more abundant in Esthwaite.

It is evident that the small differences seen in table 14 cannot play much part in bringing about the great differences in the kinds and abundance of algae and other organisms in the various lakes. The substances in fig. 29 are clearly more important but it is scarcely likely that they are responsible for all the differences, which must therefore be related either to rare elements or to organic compounds. Lund (1957) writes that culture work has shown that some algae need iron, manganese, boron, molybdenum, vanadium, cobalt and copper. Other elements are known to occur in cells. Not much work has been done, and the results of the one analysis performed on Lake District material are presented in table 15. The concentration in *Asterionella* of nickel, lead, titanium and zinc but not of silver, cobalt, chromium, molybdenum and vanadium, and the different pattern in *Staurastrum* are noteworthy, but what they indicate is not known.

Table 15

Spectrographic analyses of lake water and algae. Windermere Water, *Asterionella formosa* Hass. and *Staurastrum paradoxum* Agg. collected from Windermere in November 1950, January 1951 and August 1951 respectively. *Prasiola crispa* (Light) Menegh. from below Puffin burrows on the Farne Islands, Northumberland in July 1948. All the samples dried to a constant weight at 105°C after being washed in distilled water. Analyses of Windermere Water and *Asterionella* by Dr R. L. Mitchell of the Macaulay Institute of Soil Research and of the other algae by members of the chemistry staff of the Atomic Energy Establishment, Sellafield, Cumberland. (Lund, J.W.G. 1957. *Proc. Linn. Soc. Lond.* **167**)

| | Windermere water μg per Kg | μg per gm dry weight | | |
| | | *Asterionella formosa* | *Staurastrum paradoxum* | *Prasiola crispa* |
Element				
Ag	>0·2	>0·5	>6·0	>2·0
Co	>0·1	>0·5	>6·0	>2·0
Cr	>0·3	>20	>10	>2·0
Mo	>0·1	>5	>6·0	>2·0
Ni	0·23	83	30	>2·0
Pb	>2·0	680	10	4
Ti	>0·5	1100	60	>20
V	>0·1	>5	>6	>2·0
Zn	>8·0	1240	>15	>40

ORGANIC COMPOUNDS

Asterionella is abundant in Windermere but will not grow in the waters of the unproductive lakes unless soil extract is added. The nature of the substance which the soil contributes, and the method by which it stimulates growth, are unknown. It is thought to be an organic compound. There are many such in water, derived from the soil, from dead bodies as they decompose, and from living organisms also. Not much is known about what they are; about their function, if any, speculation outruns knowledge. It is well established that some algae produce substances toxic to others, presumably to destroy them before they can exploit the same source of nutrients. It is possible that in this all-in warfare something corresponding to the anti-missile missile has developed; certainly some of the substances produced absorb and render harmless such toxic elements as copper. Other substances facilitate the absorption of certain ions, or hold in an available form phosphate or iron, for example, which would otherwise be precipitated. Some algae grow in a light intensity too low for photosynthesis, and it has been suggested that they make use of substances which they produced and released into the water when photosynthesis was possible (Lund 1964).

Animals too are producing substances. Those investigated so far have the purpose of inhibiting the growth of other members of the same species, which presumably benefits it by removing the possibility of a population large enough to eat all the food and thereby bring about the death of every individual (Rose and Rose 1965).

These organic substances provide a rich field ripe for exploration; at the moment there is little to be said about them except to stress that as soon as a piece of water is colonized, it is altered by the products of the colonists.

CHAPTER 6

Algae

In 1909 W. and G. S. West published a list of the phytoplankton of the lakes. Between the wars W. H. Pearsall and his father made more frequent collections, published a further list, and put forward some tentative ideas about the reasons for the distribution and the periodicity which they had observed. Thereafter major advance did not start until Dr J. W. G. Lund joined the Freshwater Biological Association in 1944. He concentrated on the problem of periodicity, and made prolonged investigations of a small number of species. At the same time he gathered information on other aspects of algal ecology, working in close collaboration with the chemists and physicists, and he also cleared up a number of taxonomic points. Dr Hilda M. Canter (Mrs J. W. G. Lund) started work soon after the second war on the Chytridiales, a group of small fungi parasitizing algae, about which very little was known. Her studies, later widened to embrace some of the Protozoa which eat algae, were mainly taxonomic but she collaborated with Dr Lund in the study of periodicity. Dr J. F. Talling (1961, 1966) has recently carried out observations on algal physiology mainly in relation to light. This work is an exploration of a fundamental physiological problem rather than a contribution to an understanding of the communities found in the Lake District lakes, and it is not, therefore, described here, in accordance with the policy set out earlier. Studies of attached algae, made by Dr Maud B. Godward and Dr F. E. Round, have not passed the descriptive stage.

The Wests (1909) listed 188 species in 11 lakes and commented on differences between the lakes. They also drew attention to the periodicity of species in Windermere, the only lake from which they had regular samples. The Pearsalls (1925) took 8 or 9 samples at different times from 11 lakes, by means of a 24 mesh/cm net towed for 30 minutes. Their list includes 205 species and 5 varieties. Their table, which takes up 6 pages, shows the maximum abundance, the figure being the number of cells of each species expressed as a percentage of the cells of all species, and the constancy, which is the number of times each species was recorded expressed as a percentage of the total number of collections. A summary of these results by Pearsall (1924) is presented here in table 16 and two shorter summaries by Pearsall and Pearsall (1925) in tables 17 and 18. The dominant species are: desmids in the unproductive lakes, Wastwater and Ennerdale; *Dinobryon* in the intermediate lakes, Buttermere, Crummock and Haweswater; and diatoms in the productive lakes Derwentwater, Bassenthwaite,

72

Table 16

Dominant and constant algae in the phytoplankton of the English lakes. (Pearsall, W. H. 1924. *Rev. algol.* 1)

Dominant species, that is those which, at times, make up more than 30% of the whole phytoplankton

Rocky Lakes			Silted Lakes		
Chloroph.	Genicularia elegans	En	Melosira granulata	*Bacill.*	Wi, Es
,,	Staurastrum longispinum	En	Tabellaria fenestrata	,,	Wi, U
,,	S. jaculiferum	Wa, Bu	Asterionella gracillima		Wi, U, Es
,,	Spondylosium planum	Cr	Anabaena circinalis	*Myxoph.*	Es
,,	Sphaerocystis schroeteri	Wa, Bu	A. flos-aquae	,,	Es
	Botryococcus braunii	Cr	A. lemmermanni	,,	Wi, Es
Chrysoph.	Dinobryon cylindricum	Cr	Oscillatoria agardhii	,,	Es
	D. divergens	Bu, Cr	Aphanizomenon flos-aquae	,,	Es
Bacill.	Tabellaria fenestrata	Bu, Cr	Coelosphaerium kutzingianum	,,	Wi, Es
			Dinobryon divergens	,,	Wi, U

Constant species, that is those occurring in more than 75% of the collection from a given lake

Rocky Lakes *only*		Silted Lakes *only*	
Gonatozygon monotaenium	En, Bu, Cr	Melosira granulata	Wi, Es
Genicularia elegans	En	Cyclotella comta	Wi
Staurastrum anatinum	En, Bu	Asterionella gracillima	all
S. arctiscon	En, Cr	Coelosphaerium kutzingianum	Es
S. brasiliense var lundelli	En	Anabaena lemmermanni	Es
S. longispinium	En		
Botryococcus braunii	Bu, Cr		
Sphaerocystis schroeteri	all		
Peridinium willei	all		
Dinobryon divergens	Bu, Cr		
Surirella robusta var. splendida	En		

Constant in some rocky and some silted

Xanthidium antilopaeum	En, Cr, U
Staurastrum jaculiferum	Wa, Bu, Cr, U
S. paradoxum	Cr, Wi
Spondylosium planum	Cr, Wi, Es
Eudorina elegans	Cr, Wi, Es
Ceratium hirundinella	Cr, Wi

Table 17

The dominant species of algae in the phytoplankton of the lakes
The species of *Anabaena* are *A. circinalis* (Kütz) Hansg., *A. flos-aquae*
(Lyngb.) Bréb., and *A. lemmermanni* P. Richter.
Of five species of *Dinobryon* the commonest are *D. cylindricum* Imhof and
D. divergens Imhof. (Pearsall and Pearsall, 1925, *J. Linn. Soc. Bot.* **47**)

	Wa	En	Bu	Cr	Ha	De	Ba	Ull	Wi	Es
Asterionella formosa						d	d	d	d	d
Melosira granulata									sd	sd
Tabellaria fenestrata and var.			sd	sd	sd	sd	sd	d	sd	
Coelosphaerium kützingianum					sd				*sd	d
Oscillatoria agardhii										d
Anabaena spp.						sd			sd	sd
Dinobryon spp			d	d	*d	d				
Mougeotia elegantula							d			
Staurastrum jaculiferum	d									
„ *longispinum*		d								
Green sub-dominants	s	g	b	sp						

b = *Botryococcus braunii.*   g = *Genicularia elegans.*
s = *Sphaerocystis schroeteri.*   sp = *Spondylosium planum.*
* Present in less than 75 % of the samples.

Table 18

The constant species in the phytoplankton of the lakes
The figures show the number of species in the group that were found in
more than 75% of the collections from a lake. (Pearsall and Pearsall, 1925.
J. Linn. Soc. Bot. **47**)

	Wa	En	Bu	Cr	Ha	De	Ba	Ull	Wi	Es	Co
Desmids (Conjugatae)	1	7	3	6	12	7	4*	2	2	1	
Colonial Chlorophyceae	1		2	3	2	3			1	1	1
Diatoms (excluding *Tabellaria*)		1			1	1	1	2	3	2	1
Myxophyceae				2	1				1	3	
Peridinieae	1	1	1	2	1	1			1		1
Dinobryon			1	1		1					

* Including *Mougeotia elegantula.*

74

Algae

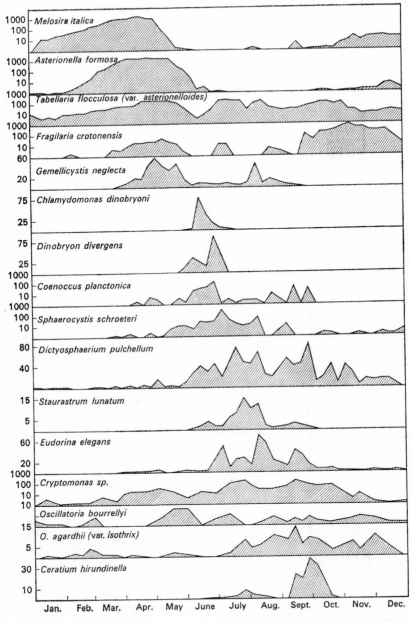

32. Succession of algae in Windermere in 1965 (data supplied by Dr J. W. G. Lund).

Ullswater, Windermere and Esthwaite. *Dinobryon* is dominant also in Derwentwater, and blue-greens in Esthwaite. A dominant species is one which at times contributes more than 66 per cent to the total number of algae in

75

the phytoplankton, a subdominant is one that contributes more than 33·66 per cent. Arranged according to the algae in this way, the lakes fall into the order arrived at on other criteria.

With Pearsall's (1932) paper the work begins to leave the descriptive and enter the explanatory stage. The establishment of a laboratory beside Windermere and the installation in it of botanists has made possible more numerous and more frequent collections than Pearsall was able to effect. Knowledge of what species occur and of when they abound has become more precise, and the explanatory stage has been carried much further. The information in Pearsall's paper has been largely superseded. In it

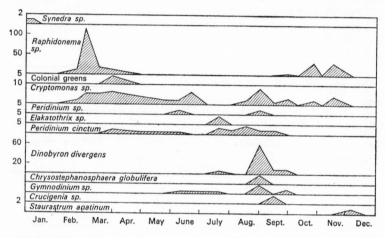

33. Succession of algae in Buttermere in 1965 (data supplied by Dr J. W. G. Lund).

there are four pages of tables showing the periodicity of the commoner species in nine lakes, but here use has been made of more recent unpublished information which Dr Lund has kindly provided (figs. 32 and 33). The year 1965 was a fairly typical one in Windermere, except in the autumn, when blue-green algae were less abundant than usual. It will be noted that the scales differ greatly from species to species, and that few species reach great abundance after the spring outburst. The summer, though not a period of high numbers, is the period of greatest diversity, and there is much variation from one year to another, a variation which has not yet been investigated with any thoroughness. The outstanding feature of the unproductive lake Buttermere (fig. 33) is the small number of cells. Both figures omit the very small cells that pass through a net. Their inclusion would reduce the difference between the lakes, but it would still be large particularly in terms of production, to which the nannoplankton does not make a big contribution because of the minute volume of each cell.

Lund started with an investigation of *Asterionella*. As a result of taxonomic studies he concluded that the species is *A. formosa* Hassall and not

76

A. gracillima (Hantsch) Heiberg., as stated by earlier workers. He showed that, not only does it not occur in the unproductive lakes, but that it will not grow in water from them unless soil extract is added. His main works on the periodicity are Lund (1949a, 1950), Canter and Lund (1948, 1953 and 1951), and Lund, Mackereth and Mortimer (1963). Lund (1949b) is a short but useful summary. Hughes and Lund (1962) present the results of culture work. Lund (1965) is a formal review of work on the ecology of algae. Lund (1964) has more of the nature of an informal discourse to a group of friends, with personal opinions and philosophical digressions. It deals mainly with his own experience.

Early in his first paper, Lund (1949a) writes 'The problem has become obscure as a result of recent work', and produces evidence that *Asterionella* is present in the open water all through the year, and that there is no need to postulate origin of the cells that produce the spring outburst from the mud, or from elsewhere in the drainage basin. Numbers are low in winter because there is little light anywhere in the lake, and because any one cell is liable to be carried from the upper illuminated zone into the dark depths, where no photosynthesis is possible. Factors that help to keep numbers low are loss down the outflow, particularly in time of flood, and parasitism. Chemical conditions in winter are favourable, and it is the physical conditions that are adverse.

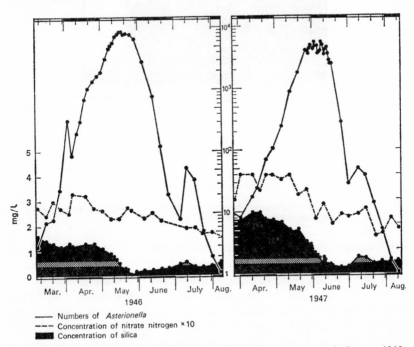

— Numbers of *Asterionella*
−−− Concentration of nitrate nitrogen × 10
▰▰ Concentration of silica

34. Numbers of *Asterionella* in Windermere, North Basin, in two typical years, 1946 and 1947 (Lund, J. W. G. (1950), *J. Ecol.* **38**).

Numbers increase in the spring mainly because the days are longer and the light penetrates more deeply into the water. The outburst starts earlier in Esthwaite because it is a shallower lake than Windermere, and therefore cells spend less time in the dark deeper regions. Increase in numbers is soon rapid and, when a big population has been produced, there is a sudden decrease. Lund observed this over a number of years and found that the

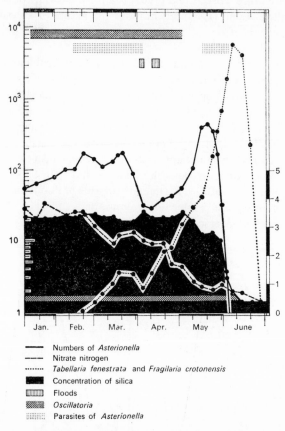

Numbers of *Asterionella*
Nitrate nitrogen
Tabellaria fenestrata and *Fragilaria crotonensis*
Concentration of silica
Floods
Oscillatoria
Parasites of *Asterionella*

35. Numbers of *Asterionella* in Esthwaite in 1949 (see also fig. 34) (Lund, J. W. G. (1950), *J. Ecol.* **38**).

one feature that almost always coincided with the decrease was a fall in the concentration of silica to between 0·4 and 0·6 mg/l. In his second main paper (Lund 1950) he presents graphs for Esthwaite and for the two basins of Windermere for the years 1945 to 1949 inclusive. Fig. 34 has been selected from these fifteen for reproduction here. All the others except two (figs. 35 and 36), though not identical, are similar. In 1964 he produces figures for the years 1945–60 (fig. 37). The association of the rapid decline of the alga and the decrease in the concentration of silica to about 0·5 mg/l is so regular that there can be little doubt that the one is the cause of the other. It would seem that actively growing cells cannot survive when some essential requirement reaches a level at which it cannot be taken from the water.

In Esthwaite in 1949 (fig. 35) numbers increased more slowly than in other years, partly owing to parasitism and partly owing to greater numbers than usual of *Oscillatoria agardhii* var *isothrix*. This alga is

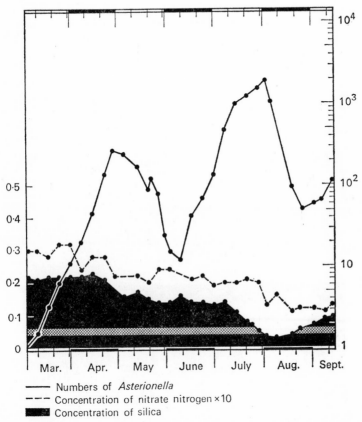

——— Numbers of *Asterionella*
- - - Concentration of nitrate nitrogen ×10
▓▓ Concentration of silica

36. Numbers of *Asterionella* in Windermere in 1948 (see also fig. 34) (Lund, J. W. G. (1950), *J. Ecol.* **38**).

believed to have affected *Asterionella* adversely by taking the light, not by means of a toxic secretion, a point that will be returned to later. Numbers of both algae were reduced by floods in April after which *Oscillatorai* was severely parasitized and *Asterionella* was not. Severe infestation then returned to check growth. Meanwhile *Tabellaria* and *Fragilaria* had increased and they reduced the concentration of silicate to the critical level. The outcome was a final population of *Asterionella* less than one-tenth of the usual maximum.

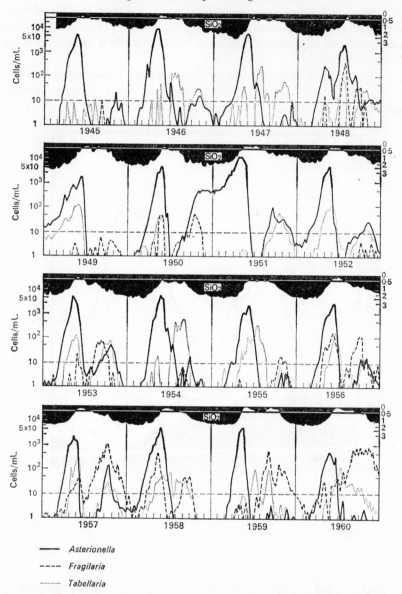

——— Asterionella

---- Fragilaria

·········· Tabellaria

37. The periodicity of *Asterionella formosa* Hass, *Fragilaria crotonensis* Kitton, *Tabellaria flocculosa* (Ehr) Grun. var. *asterionelloides* (Grun.) Knuds. and the fluctuations in the concentration of dissolved silica, in the 0–5 m. water column of the northern basin of Windermere from 1945 to 1960 inclusive (Lund, J. W. G. (1964), *Ver. int. Verh. Limnol*, **15**).

In Windermere North Basin in 1948 (fig. 36) there was a decrease in the numbers of *Asterionella* in May, coinciding with a period of dry weather. Increase in numbers was resumed when rain fell. It is postulated that some auxiliary substance was in short supply during the drought and that the supply of it was replenished by the rain. Evidently, since there was no comparable check in Esthwaite or in the South Basin, into which Esthwaite discharges, there was no deficiency of the auxiliary substance there.

In high concentrations of phosphate, the limiting threshold of silicate is reduced, by an amount which depends on illumination and temperature. Most complete use of silicate is made when the light is not too bright and the temperature low, and cultures under these conditions last longer and produce more cells (Hughes and Lund 1962).

The main school in opposition to those who explain rise and fall of algae in terms of physical and chemical factors attributes the success of a species to the production of a toxin that eliminates rivals, and its ultimate decline to the poisonous effects of its own products. The existence of such toxic products is well attested, but Lund has not found it necessary to invoke them to explain the course of events in the English lakes. In the autumn blue-greens reach a maximum and, as they belong to a group noted for its toxins, it might be expected that the numbers of other species would be low then, if toxins are a controlling factor. In fact they are not low, and it is a time when *Asterionella* begins to increase again. Moreover, cultures of *Asterionella formosa* and *Fragilaria crotonensis* do not show any effect of either upon the other (Talling 1957).

After the spring outburst, numbers of *Asterionella* generally remain low all through the summer. Usually the concentration of silica is low too but this is not invariable, and there is probably some other substance in short supply holding *Asterionella* in check. Other species rise and fall after *Asterionella* has passed its peak, though the number of cells never reaches the total attained in spring. Presumably these species each have different requirements of the inorganic and organic substances in the water, but the field is unexplored.

Very small algae are eaten by Rotifera and Crustacea but larger ones are too big for these animals to tackle. The one exception is the rotifer, *Asplancha*, which can ingest large algae, including *Asterionella*, though there is no reason to believe that it does so in sufficient numbers to affect the size of the whole population. Recently Canter and Lund (1968) have observed that certain Protozoa, other than Ciliates, attack large algae, and they draw attention to the paradox that it is the smallest animals that prey on the large algae whereas the largest members of the zooplankton eat only the small algae. This line of work is hampered, as pioneer ecological work often is, by taxonomic difficulties, and the Lunds attempt no more than a tentative identification of genera. It remains to be discovered whether each algal species is preyed on by a distinct protozoan species. Cells of *Sphaerocystis* are engulfed whole, the typical rhizopod method of feeding; cells of *Dictyosphaerium* are pierced by a filopodium and the contents are digested outside the amoeboid body. Populations of algae that are infested by one of these organisms may be reduced considerably within a week or two, but the work has not yet reached a stage where it is possible to determine how important a part they play in the normal periodicity of any alga.

In the autumn the numbers of *Asterionella* rise again but deteriorating physical conditions prevent the attainment of a peak as large as the spring one.

Lund (1959b) draws attention to the significance of sinking. In still water it is the only way in which an alga can pass out of a zone in which it has exhausted some substance that it requires to one in which there may be more. There is a danger that it may be carried out of the photic zone before turbulence carries it up again, but to set against this there are advantages in being in dim light when silicate is exhausted; it appears to be the actively increasing cells that die, and the slowly increasing ones, in places where conditions are unfavourable, that survive.

A second species on which Lund has worked is *Melosira italica* subsp. *subarctica*. He describes observations extending over a period of seven and a half years (Lund 1954) and then more frequent observations during a short period (Lund 1955). *Melosira* occurs in the open water in winter, reaches a maximum before *Asterionella,* and then disappears. Cells are always to be found on the bottom. This alga has two peculiarities. The first is a rapid rate of sinking. Lund (1959) compares its rates with those of *Asterionella* in a table of which the most striking feature is perhaps the size of the range of the rate of each species. Whatever the reason for this, the fastest rate for *Melosira* is twice that of the fastest *Asterionella* and the slowest *Melosira* covered the measured distance in a time only one-eighth of that taken by the slowest *Asterionella*. The average difference lies somewhere between these two figures. The other peculiarity of *Melosira* is its ability to survive on or in the mud, even anaerobic mud, for several years.

When the thermocline is so deep that circulating water can sweep cells from the mud in deep water up to the surface, gales result in the appearance of *Melosira* in the water. Numbers are sometimes so high so suddenly that development from a small population in the water could not have produced them in the time. Further, Lund was able to demonstrate a decrease in the number on the bottom corresponding to an increase in the water. Having attacked other workers for suggesting a similar origin for *Asterionella*, he did not make any statement about *Melosira* until the evidence was strong. Transferred from the mud to the water, the cells start to increase but, after a time, the short days and low light intensity, with other unfavourable factors in winter, bring the increase to a halt. Floods may reduce numbers and, if ice forms and lasts for a few days, the entire population sinks to the bottom. The main time of increase is the spring and it appears to come to an end at a concentration of silicate a little higher than that which limits the increase of *Asterionella*. The rate of sinking is so rapid that it is rarely long, once a lake has stratified, before the epilimnion is still enough for a sufficient length of time for all the cells to fall out of it.

Melosira is parasitized by a chytrid and eaten by *Asplanchna* but neither organism appears to affect its numbers severely.

A strong gale bringing cells up from the bottom at the same time as heavy rain washing them out of the lake are unfavourable to *Melosira*. The loss will be greater in shallow lakes and in lakes that are small relative to the catchment area. A gale without heavy rain may favour the species by bringing more cells into circulation. If unusually strong, it may uncover and bring up cells that fell to the bottom two or three years previously. In the Lake District, ice is likely to form at a time of year when *Melosira* cells in the water are doing no more than holding their own numerically, and therefore their fall to the bottom is not an adverse event, provided that, when the ice melts, the wind is strong enough to bring them back into the water. The conclusion is that the numbers attained by *Melosira* are sometimes determined by physical factors alone.

In Esthwaite in 1949 circumstances unfavourable to *Asterionella* enabled *Tabellaria* and *Fragilaria* to exceed it in abundance (fig. 35). Lund (1964) answers the question why *Asterionella* is almost always the winner in what must be a race for the available nutrients. Cultured with it, *Asterionella* divides more often than *Tabellaria*, and in this way soon forms a much more numerous population. *Fragilaria's* rate of increase, on the other hand, is as fast as that of *Asterionella*, and its lack of success is due to lower numbers at the start of the race. Why its numbers fall so much below those of *Asterionella* during the winter is not known.

Both these algae frequently reach small peaks after *Asterionella*, and presumably must be able to take up silicate, and perhaps phosphate too, at lower levels than *Asterionella* can.

Although chemical conditions are not good during the summer, more species contribute to algal production during the period June to October than during the other seven months of the year. It is difficult to separate the effects of the two main physical variables, but temperature is undoubtedly of considerable importance. Lund (1964) records that, of twenty-five species of algae from Windermere tested in culture, every one grew fastest when the temperature was between 18 and 25°C, and rebuts the idea that the attainment of maximum numbers in spring is connected with a low optimum temperature.

Lund (1957, 1961, 1964) turns to the question of numbers and productivity. The algae of the phytoplankton are unusual among organisms in that the larger ones at least are not eaten in significant quantity, and in that many of them increase until deficiency of some essential requirement brings further multiplication to an end. It is, therefore, justifiable to assess

Table 19

Approximate dry weights of the largest crops of algae in 5 cubic metres (surface −5 m deep). Only those crops weighing more than 0·5 g/m² are included and only those species whose weight is known. Anabaena spp., often abundant in Esthwaite, is omitted because the number of spores made the weight unreliable. An average of 6 μg per 10^6 cells has been taken for coccoid μ-green algae and of 16 μg per 10^6 cells for Stichococcus spp. (Lund, J. W. G. 1961. Verh. int. Ver. Limnol. 14)

| | Windermere | | | | | | Blelham Tarn | | | Esthwaite | | |
| | North Basin | | | South Basin | | | | | | | | |
	1956	1957	1958	1956	1957	1958	1956	1957	1958	1956	1957	1958
Asterionella formosa	4·50	3·45	9·18	3·10	11·68	1·57	1·54	11·61	21·10	2·59	7·38	3·88
Cyclotella glomerata	3·10	1·09	0·98	14·60	0·85	0·73	–	–	–	–	–	–
Dictyosphaerium ehrenbergianum	0·52	–	–	1·92	0·94	– *	–	–	–	–	–	–
Fragilaria crotonensis	–	1·65	0·71	2·98	1·52	3·88	–	–	–	–	1·25	–
Melosira italica subsp. subarctica	–	–	–	3·28	0·61	6·28	7·81	4·64	2·67	2·56	7·35	16·46
Tabellaria flocculosa var. asterionelloides	–	–	–	2·02	0·65	–	0·81	–	–	7·00	1·64	2·27
Oscillatoria agardhi var. isothrix	–	–	–	–	1·69	2·01	–	–	–	4·16	6·11	3·77
Coccoid μ-green algae and Stichococcus	–	–	–	–	1·29	–	–	1·51	1·70	–	–	0·91
Eudorina elegans	–	–	–	–	–	1·54	–	–	–	–	–	–
Aphanizomenon flos-aquae	–	–	–	–	–	–	0·97	–	4·28	2·11	0·97	6·46

* D. pulchellum 0·64.

the productivity of a lake in terms of the maximum numbers reached by the various species found in it, a process which, in other biotopes, can lead to misleading conclusions.

Numbers alone cannot be used to measure productivity or production, without reference to the size of the various species involved. Lund (1961) gives a table showing the dry weights of the largest crops under one square

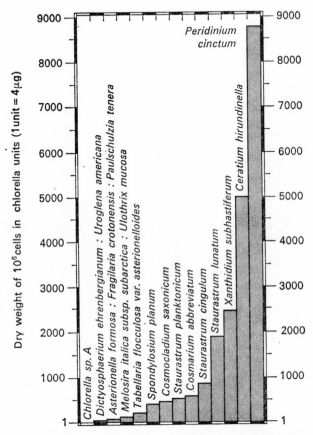

38. Average dry weights of certain plankton algae of Windermere based on estimates from natural or cultivated populations (Lund, J. W. G. (1964), *Ver. int. Verh. Limnol.* **15**).

metre of lake surface in the two basins of Windermere and in Esthwaite, and it is reproduced here, in a slightly different form, in table 19. He also gives the dry weight of some individual species, data which he presents in the form of a figure in a later paper (1964) (fig. 38). *Chlorella* is one of the smallest of the μ-algae and larger algae may be up to 9000 times as heavy. In well silicified diatoms carbon amounts to about 20 per cent of the dry weight, in other algae to about 40 per cent. The proportion of carbon in the organic matter, that is in the dry weight less the weight of the ash, is similar in all algae.

The periodicity of the μ-algae is described and illustrated by Lund (1961). They tend to be most abundant in spring and scarce in winter.

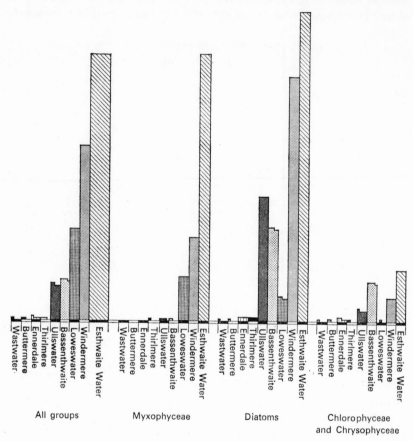

All groups   Myxophyceae   Diatoms   Chlorophyceae and Chrysophyceae

Note

▫ = Basic areal unit

39. Maximum standing crops of algae expressed as ratios in nine lakes (Lund, J. W. G. (1957), *Proc. Linn. Soc. Lond.* **167**).

Lund is of the opinion that they contribute less to total production than larger algae but is not certain about this because their populations are subject to unknown and possibly large losses from predation.

The contribution of the μ-algae being uncertain, and the dry weights of some species being unknown, it is not possible to compare production in lakes in terms of the weight of the maximum crop of each species. It has, however, been done in terms of maximum numbers (Lund 1957) (fig. 39).

The crop in the most productive lakes is over a thousand times as great as in the least productive; it is clearly unrelated to the concentration of the major inorganic ions. Windermere South Basin, in which the hypolimnion does not become deoxygenated, comes between Esthwaite and Loweswater, in both of which it does, an observation that supports the earlier contention that there is not a fundamental difference between the two types of lake.

The only attached algae that have been studied in detail belong to the genus *Tabellaria*, in which Knudson (1954) found four species: *T. quadrisepta* in tarns and pools; *T. fenestrata* in tarns and lakes; the widely distributed *T. flocculosa*; and the rare *T. binalis*. *T. flocculosa* var. *flocculosa* can be attached or planktonic but var. *asterionelloides*, already mentioned, is truly planktonic. Knudson was also able to distinguish various strains, some of them confined to the lakes of a single drainage area.

The winter population of *Tabellaria* fluctuates little, though the level varies considerably in different years. There is an increase during the months of February, March and April which is brought to an end by, Knudson (1957) suggests, deficiency of some essential nutrient. Trends later in the summer are variable.

Cannon, Lund and Sieminska (1961) suspended cultures of *T. f. flocculosa* in bottles at different depths. Each bottle was provided with what was believed to be sufficient nutrients and, at the end of a week, the contents were counted. Results were expressed as the number of divisions (fig. 40). During the first four months of the year the best growth never amounted to as much as two divisions per week, and evidently temperature was the governing factor. Thereafter the number of divisions increased and, as a peak was reached well after the longest day, it was evident that temperature was still the paramount factor. This was true not only at 0·5 m, as shown in fig. 40, but down to about 3 m. There was more than enough light in these upper layers, but in deeper water there was not, and rate of division was symmetrical about a line indicating the longest day. There was no evidence of inhibition of photosynthesis by light in summer, but at low temperatures, in winter or early spring, there was sometimes less increase at 0·5 m than at 1 or 2 m. Increase took place down to a depth of 8 m in April, May, June and July, to 6 m in August and September, and to 4 m in November.

The authors discuss the summer minimum observed by Knudson. They conclude that it was not caused by inhibition, because other algae are at a minimum at that time. The effects of dislodgement by wave action and predation are less certain; but the first is likely to be too irregular to produce what appears to be a regular phenomenon, and, in the light of present knowledge, there is no reason to believe that there is a great increase in the number of animals grazing on epiphytes in summer. They agree, accordingly, with Knudson that depletion of some essential substance is the most likely cause.

The species of attached algae have been recorded by Godward and by Round. Godward (1937) was primarily interested in the differences between the different substrata in Windermere, and the number of her stations on any one substratum was small. The work of Pearsall on the rooted vegetation, and studies of the distribution of animals, suggest that, on any one type of substratum, conditions are different in different parts of the lake. Godward's work cannot, therefore, be taken as more than

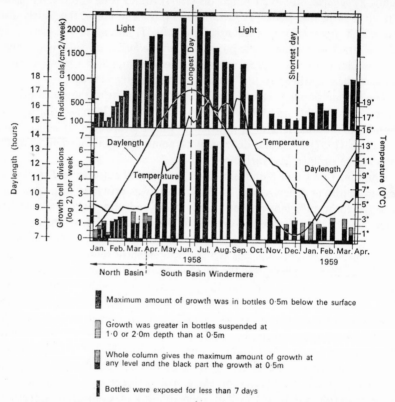

Maximum amount of growth was in bottles 0·5m below the surface

Growth was greater in bottles suspended at 1·0 or 2·0m depth than at 0·5m

Whole column gives the maximum amount of growth at any level and the black part the growth at 0·5m

Bottles were exposed for less than 7 days

40. The maximum amount of growth of *Tabellaria flocculosa* var. *flocculosa* in bottles suspended in Windermere. Growth is expressed in the number of synchronous divisions (i.e. $\log_2$ increments) per week. The amount of growth is shown by the histograms at the base of the figure (Cannon, D., Lund, J. W. G. and Sieminska, J. (1961), *J. Ecol.* **49**).

a preliminary survey. There are 72 pages of it, and the main results are summarized here in table 20, a condensation which inevitably involves considerable approximation of the extensive data on depth and seasonal distribution. Round confined his attention to the algae on the sediments, and his work too was of a preliminary nature, a great deal of ground being covered superficially. He visited seventeen lakes and collected at two stations in Brothers Water and Ullswater and at three stations in Windermere and Blelham. He divides (Round 1957b) the lakes into two groups on the basis of chemical analysis of the sediments as follows:

Unproductive group	*Productive group*
Crummock	Windermere (3)
Wastwater	Rydal
Ennerdale	Coniston
Derwentwater	Elterwater

Biological Studies of the English Lakes

Brothers Water (2)	Ullswater N. Basin
Bassenthwaite	Blelham
Loweswater	Loughrigg Tarn
Ullswater S. Basin	Esthwaite
Buttermere	Grasmere

It will be noted that Loweswater is placed in the unproductive group, although all other studies have shown it to be one of the more productive lakes. Moreover, two stations in Ullswater come into different groups, and it is the South Basin that is classified as unproductive, and the North as productive, the reverse of what is likely. There is little doubt, therefore, that an investigation in which many stations are established in each lake will modify the preliminary picture.

Round (1957b) gives a list of Bacillariophyceae occurring on littoral sediments, which covers 11 pages and includes more than 300 species. It includes algae that are planktonic as well, and in a second paper (Round 1957c), the significant part of whose title appears to be 'bottom-living algae', the total is much lower. A third paper (Round 1957e) is devoted to groups other than Bacillariophyceae.

These are sometimes outnumbered by the Cyanophyceae, but in general the other groups are scarcer both in number of species and in number of individuals. Round's findings are summarized in table 21. The Bacillariophyceae (Round 1960) usually reach highest numbers in spring, though sometimes not till summer or even autumn. There is variation with season and place, and no general conclusions appear to be justified. The life-histories of the Cyanophyceae (Round 1961a) show generally, though not invariably, a peak of abundance in spring coinciding with that of the planktonic algae of other groups; only rarely was the peak in late summer as is common with the planktonic algae of the same group.

Optimum growth of this epipelic community is at about 3 m and it is vigorous also at 1 and 5. Why there should be less at 2 and 4 m is still unknown. There is little growth below 8 m in Windermere and Blelham (Round 1961b).

Distribution of attached algae in Windermere (Godward 1937). The species are arranged according to their distribution on three types of substratum, at seven depths (m), and in time. Sp = spring, su = summer, au = autumn, wi = winter, ayr = all year round. If there is no indication of season except ayr, the species is present at all times with no maximum

Species	hard sub-stratum	living plants	spray zone	0-0·5	0·5-1	1-2	2-3	3-3·5	3·5-4	4-6	sp	su	au	wi	ayr
Pleurocapsa fusca	X		X	X								X			X
Tolypothrix distorta var. penicillata	X		X												
Phormidium autumnale	X		X	X							X	X			X
Calothrix parietina	X		X	X							X				X
Schizothrix funalis	X		X	X								X			X
Pleurocapsa fluviatilis	X	X													
Nostoc sp.	X		X												
Synedra radians	X			X	X	X					X		X	X	
Achnanthes minutissima	X	X		X	X	X	X				X		X	X	
Achnanthes microcephala	X	X		X	X	X	X							X	X
Ulothrix zonata	X	X		X							X				
Cymbella ventricosa	X			X	X						X				
Phormidium foveolarum	X			X								X			
Homoeothrix fusca	X			X											
Dichothrix orsiniana	X			X											
Phormidium laminosum	X			X											
Draparnaldia plumosa	X	X		X							X			X	
Stigeoclonium amoenum	X	X		X							X		X	X	
Oedogonium spp.	X			X	X							X			
Spirogyra sp.	X	X		X	X	X	X						X		
Mougeotia sp.	X	X		X	X	X	X								
Zygnema sp.	X	X		X	X	X	X								
Nostoc verrucosum	X			X											X

Dichothrix baueriana var. *crassa*	X				X	X							X
Denticula tenuis	X				X	X				X			
Plectoma tomasiniana var. *cincinnatum*	X				X	X						X	
Cladophora spp.	X				X	X							
Sphaerobotrys fluviatilis		X	X									X	
Hormidium subtile		X	X										
Hyalotheca mucosa		X	X							X			
Chaetophora sp.		X	X										
Naegelellia britannica		X	X									X	
Chaetopeltis orbicularis		X	X							X			
Batrachospermum moniliforme		X	X						X		X		
Coleochaete pulvinata		X	X								X		
Clastidium setigerum		X	X							X			
Gloeotrichia sp.		X	X										
Mougeotia parvula		X	X										
Lyngbya perelegans		X	X							X			X
Stigeoclonium farctum		X		X	X					X			
Ulvella frequens		X		X	X					X			
Bulbochaete intermedia		X		X	X					X			
Coleochaete scutata		X		X	X						X		
Coleochaete soluta		X		X	X						X		
Tabellaria flocculosa		X		X	X								
Tabellaria fenestrata		X		X	X								
Navicula subhamulata		X		X	X						X		
Gomphonema olivaceum		X		X	X				X		X		
Coleochaete nitellarum		X				X	X	X					
Eunotia veneris		X				X	X	X				X	
Eunotia lunaris		X				X	X	X					
Cocconeis placentula		X				X	X	X				X	
Epithemia turgida		X				X	X	X					
Chamaesiphon cylindricus		X				X	X	X					X
Chamaesiphon confervicola		X				X	X	X					
Lyngbya purpurascens		X				X	X	X					

Table 20b

Distribution of attached algae on Windermere (Godward) 1937
On dead leaves and debris

0–8 m	0–10 m	0–12 m
Navicula sp.	*Navicula radiosa*	*Pinnularia dactylus*
Frustulia rhomboides	*N. rhynchocephala*	
Nitzschia palea	*Pinnularia gibba*	
N. angustata	*P. mesolepta*	
N. sigma	*P. acrosphaerica*	
Surirella robusta	*Stauroneis anceps*	

Table 21

The commoner bottom-living algae (Round 1957d, e)

1. Species abundant in lakes of both types

Caloneis silicula	*Nitzschia dissipata*	*Pleurotaenium ehrenbergii*
Neidium iridis	*N. palea*	*Micrasterias rotata*
N. bisculcatum	*N. sigmoidea*	*Chroococcus turgidus*
Stauroneis phoenicenteron	*Surirella robusta*	*Merismopedia glauca*
S. anceps	*Neidium affine*	*Anabaena constricta*
Navicula pupula	*N. dubium*	*Oscillatoria limosa*
N. cryptocephala	*Pinnularia microstauron*	*O. splendida*
N. rhynchocephala	*Cymbella prostrata*	*O. bornetii*
N. radiosa	*Hemidinium nasutum*	*O. irrigua*
Pinnularia mesolepta	*Peridinium willei*	*Pseudanabaena catenata*
P. divergens	*Euglena mutabilis*	
P. major	*Euastrum ansatum*	
P. viridis	*E. verrucosum*	
P. gibba	*E. oblongum*	

2. Species almost confined to unproductive lakes

Eucocconeis flexella	*Cymbella ventricosa*	*Navicula cuspidata*
Frustulia rhomboides	*Gomphonema acuminatum*	

3. Species confined to or almost confined to productive lakes

Caloneis amphisbaena	*Navicula pupula*	*Pinnularia polyconca*
Neidium hitchcockii	*N. bacillum*	*P. cardinaliculis*
Diploneis fennica	*N. hungarica*	*Nitzchia acicularis*
D. ovalis	*N. placentula*	*Surirella moelleriana*
Stauroneis smithii	*N. menisculus*	*S. gracilis*
		Amphora ovalis var. *pediculus*

4. Species more frequent and abundant in unproductive lakes

Gomphonema acuminatum   *Sternopterobia intermedia*   *Netrium digitus*
G. constrictum   *Surirella biseriata*
Nitzschia ignorata

5. Species more frequent and abundant in productive lakes

Gyrosigma acuminatum   *Phacus pleuronectes*   *Scenedesmus quadricauda*
Stauroneis anceps   *pyrum*   *Closterium ehrenbergii*
Pinnularia subsolaris   *Trachelomonas volvocina*   *Spirotaenia condensata*
Cymbella naviculiformis   *T. hispida*   *Aphanothece stagnina*
ehrenbergii   *Cryptomonas ovata*   *Synechococcus aeruginosus*
Euglena viridis   *C. erosa*   *Merismopedia elegans*
E. spirogyra   *Synura uvella*   *Holopedia geminata*
Gymnodinium aeruginosum   *Pediastrum boryanum*

CHAPTER 7

Zooplankton

Until the recent publication by Smyly (1968) on the planktonic Crustacea of the lakes, studies of zooplankton from within the laboratory had been concerned with problems of general interest in no way peculiar to Windermere or to the English lakes. Outside investigators had amassed a considerable amount of qualitative information. That relating to Cladocera and Copepoda is brought together in the Freshwater Biological Association's Scientific Publications on these two groups (Scourfield and Harding 1958, and Harding and Smith 1960). Fortnightly collections extending over fifteen months were sent to G. H. Wailes for examination, and he visited the laboratory in August 1937 and 1938. Regular collections have also been sent to A. L. Galliford, who has reported on the Rotifera.

Table 22 shows the distribution of the truly planktonic species in the lakes and fig. 41 shows the mean standing crops. Table 23 shows the species caught in deep water near the mud surface. There is more plankton in the productive than in the unproductive lakes (fig. 41) but no significant difference in the species found can be observed. Of the Cladocera, *Daphnia hyalina*, *Bosmina coregoni* var. *obtusirostris*, *Leptodora kindti* and *Bythotrephes longimanus* occur in most of the lakes; *B. longirostris* is confined to the Windermere drainage area and *Ceriodaphnia* spp. to Windermere and Esthwaite; *Diaphanosoma brachyurum* and *Holopedium gibberum* are distributed in an irregular way which nobody at present has been able to explain. *Holopedium* occurs in Grasmere and Rydal, two small lakes receiving the sewage from a moderate resident and a large summer-migrant population, but it has never established itself in Windermere into which these two lakes flow. In contrast the species is generally present in small numbers in Ullswater, where it may be able to survive only because of continual reinforcements from Brothers Water, another small lake in which numbers are high. *Holopedium* did occur in Blelham Tarn until the amount of sewage entering the water was increased, but, in view of its continued occurrence in the sewage-enriched Grasmere and Rydal, a connexion between the events is far from certain. At the other end of the series *Holopedium* occurs in Ennerdale and in that lake alone.

Of the Copepoda, *Diaptomus gracilis* occurs in every lake and is generally the most numerous; indeed it is generally the most numerous of all the zooplankton Crustacea. *D. laticeps* has been recorded in Haweswater and Windermere. *Cyclops leuckarti* is a species of tropical and temperate latitudes nearing its northern limit of distribution in the Lake District,

94

Table 22

Composition of the planktonic Crustacea. (Smyly, W. J. P., 1968. *J. Anim. Ecol.* **37**)

Lake group:	I						II		III			IIIa				IV		
Lake:	Wastwater	Ennerdale	Buttermere	Crummock Water	Thirlmere	Haweswater	Derwentwater	Bassenthwaite Lake	Coniston Water	Ullswater	Windermere*	Brothers Water	Elterwater	Grasmere	Rydal Water	Blelham Tarn	Esthwaite Water	Lowes water
CLADOCERA																		
Diaphanosoma brachyurum	–	–	–	–	+	G	+	+	–	–	+	+	+	+	+	–	+	–
Holopedium gibberum	–	+	–	–	–	–	–	–	–	+	–	+	–	N	+	–	–	–
Daphnia hyalina	+	–	+	+	+	+	+	+	+	+	+	+	+	+	+	+	+	+
Ceriodaphnia spp.	–	–	–	–	–	–	–	–	–	–	–	–	–	–	+	–	+	–
Bosmina coregoni var. *obtusirostris*	+	+	N	N	+	+	+	N	+	+	+	+	+	+	+	+	+	+
B. longirostris	+	–	–	–	–	N	–	–	+	+	+	–	–	+	+	–	+	–
Leptodora kindti	+	–	N	N	+	N	+	+	+	N	+	+	+	+	+	+	+	+
Bythotrephes longimanus	+	+	–	+	+	+	+	+	+	+	+	–	–	–	+	–	–	–
COPEPODA																		
Diaptomus gracilis	+	+	+	+	+	+	+	+	+	+	+	+	+	+	+	+	+	+
D. laticeps	–	–	–	–	–	+	–	–	–	–	+	–	–	–	–	–	–	–
Cyclops leuckarti	–	–	–	–	–	–	+	+	–	–	+	+	–	–	+	+	N	+
C. strenuus	+	–	+	+	+	+	–	–	+	–	+	+	–	+	+	–	–	+
Limnocalanus macrurus	–	G	–	–	–	+	–	–	–	–	–	–	–	–	–	–	–	–
ROTIFERA																		
Asplanchna priodonta	+	+	–	+	+	+	–	–	+	+	+	+	+	–	+	+	+	+

N, New records; G, unconfirmed records of Gurney (1923).

* Data from Gurney (1923) and Wailes (1939).

Table 23

Average number of each species found in the Jenkin surface-mud samples taken in deep water (no./38·5 cm²). (Smyly, W. J. P. 1968. *J. Anim Ecol.* **37**)

	Wastwater	Ennerdale	Buttermere	Crummock Water	Thirlmere	Haweswater	Derwentwater	Bassenthwaite Lake	Coniston Water	Ullswater	Brothers Water	Rydal Water	Elterwater	Grasmere	Belham Tarn	Loweswater	Esthwaite Water	Frequency of occurrence
COPEPODA																		
Diaptomus gracilis	1	—	—	—	—	—	—	—	1	—	—	—	3	—	3	—	2	3
Cyclops agilis	—	—	—	—	—	—	—	1	—	—	—	—	1	—	1	—	12	4
C. albidus	—	—	—	—	—	—	1	—	—	—	—	—	—	1	—	—	—	4
C. bicuspidatus	12	4	3	4	7	3	—	11	8	—	2	1	5	—	4	—	24	17
C. fimbriatus	—	—	—	—	—	—	2	4	—	1	—	25	12	—	4	1	4	6
C. leukarti	—	—	1	2	—	3	40	—	—	—	70	217	—	179	3	—	91	8
C. strenuus	—	3	5	3	9	4	1	6	1	—	2	17	2	12	2	49	—	17
C. viridis	3	3	—	—	—	—	—	—	—	5	—	1	—	1	2	10	4	6
C. dybowskii	—	—	—	—	—	—	—	—	2	2	9	—	3	—	—	—	1	3
Canthocamptus staphylinus	—	—	—	3	—	—	—	—	—	—	—	—	—	3	—	4	1	3
Copepod nauplii	—	—	—	—	—	—	2	—	—	2	—	2	—	—	—	—	1	1
CLADOCERA																		
Latona setifera	—	—	1	—	—	—	—	1	—	—	—	—	—	—	—	—	—	1
Diaphanosoma brachyurum	—	—	—	—	—	—	—	1	—	—	—	—	—	—	—	—	1	1
Holopedium gibberum	—	—	1	1	—	—	1	—	—	1	—	—	1	4	2	1	—	6
Daphnia hyalina	—	—	1	—	—	—	—	1	—	—	—	—	—	—	—	10	5	2
Simocephalus vetulus	—	—	—	—	—	—	—	—	—	—	—	—	—	—	—	—	—	1
Ceriodaphnia sp.	—	—	—	—	—	1	—	—	1	—	9	21	1	3	—	—	187	10
Bosmina sp.	4	6	10	—	—	—	1	1	1	—	14	1	4	3	—	—	1	10
Ilyocryptus sordidus	4	2	1	2	—	—	2	—	—	—	1	—	1	—	—	1	—	5
I. acutifrons	—	—	—	—	4	—	4	—	—	—	—	—	—	—	—	—	—	2
Eurycercus lamellatus	—	—	5	1	—	—	2	—	—	—	—	—	—	—	—	—	—	4
Monospilus dispar	—	—	3	9	6	—	1	—	1	1	2	—	3	—	—	—	1	10
Alona affinis	—	—	3	—	1	1	—	1	—	—	—	1	—	42	—	—	1	2
A. guttata	6	4	1	1	—	—	1	—	—	—	1	—	1	1	—	—	—	4
A. quadrangularis	—	—	3	—	1	—	—	—	—	—	—	—	—	—	—	—	—	4
A. rectangula	—	—	1	—	12	—	—	—	—	—	—	—	1	—	—	1	—	1
Alonella nana	—	4	1	1	—	—	—	—	—	—	—	—	—	—	—	—	—	10
Chydorus sphaericus	3	4	7	4	—	—	2	2	26	—	—	1	1	—	—	—	—	3
C. piger																		8
Leptodora kindti	—	—	—	—	—	—	1	—	—	—	—	—	—	—	—	—	—	1

96

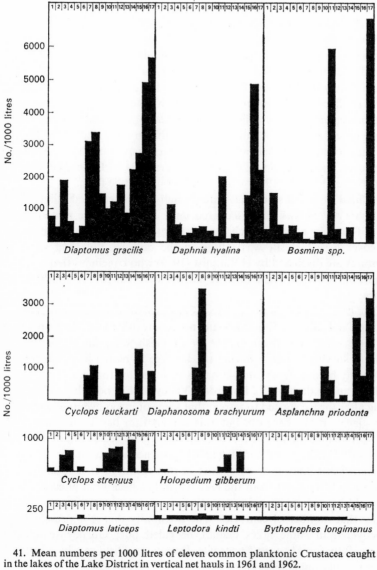

41. Mean numbers per 1000 litres of eleven common planktonic Crustacea caught in the lakes of the Lake District in vertical net hauls in 1961 and 1962.

1. Wastwater
2. Ennerdale
3. Buttermere
4. Crummock Water
5. Thirlmere
6. Haweswater
7. Derwentwater
8. Bassenthwaite Lake

9. Coniston Water
10. Ullswater
11. Brothers Water
12. Rydal Water
13. Elterwater
14. Grasmere
15. Blelham Tarn
16. Loweswater

17. Esthwaite Water

(Smyly, W. J. P. (1968), *J. Anim. Ecol.* **37**).

whereas *C. strenuus* is a species of arctic and temperate regions. *C. leuckarti* was taken in seven lakes, and in four of these it occurred alone. In only one lake, Ennerdale, was neither species taken.

Smyly finds that the plankton of the lakes has not changed since Gurney sampled it forty years ago and he quotes other evidence that it is a stable community. It is therefore hard for a new species to establish itself. Smyly believes that this stability and occasional chance happenings have produced the peculiar features which several lakes show today.

Limnocalanus macrurus and *Mysis relicta*, both found in Ennerdale but in no other lake, have attracted considerable interest since their discovery. Gurney records that Pearsall found *Limnocalanus* to be the most abundant species in the plankton of the deep eastern part of Ennerdale Water but Harding and Smith (1960) write that it could not be found in 1956. Smyly has failed to find it since that date though he has searched diligently.

Mysis relicta was not discovered until 1941 when the remains of one were found in a cone which Dr Pennington had suspended in order to discover how fast silt was falling to the bottom. This and subsequent captures are discussed by Holmquist (1959) who concludes that the species is not numerous in Ennerdale. It has not been caught during the last few years.

The discovery of these two species has provoked speculation about how they gained access to Ennerdale and not to any other lake. Both are thought to have originated from marine or brackish-water ancestors that were able to survive isolation during the Ice Age in basins in which the water gradually became fresh later. It has been suggested that sea water was held behind an ice dam in the Ennerdale Valley, but in no other valley, for so long after the ice started to retreat that some of it was eventually trapped in the glacial basin beneath, but this explanation has been challenged. It is possible that the problem is not one of original access but of subsequent survival. Harding and Smith (1960) have no hesitation in writing that *Limnocalanus* 'may now be extinct in this country'. If a species can disappear in forty years from the lake which has been least affected by man, it could have disappeared from others in a period covering three or four times that number of centuries. The possibility that it is present but undiscovered in some lakes cannot be ruled out; Gurney searched unsuccessfully for *Mysis* in Ennerdale. It remains a field in which little more than speculation is possible.

Table 24 shows the Rotifera recorded. It is based on the table in Wailes (1939), which itself is taken largely from the records of Bryce, and on the records of Galliford (1947, 1949).

Wailes (1939) records the following Protozoa, but, as most of the animals in this group are not recognizable in the preserved state, he found mainly only the species present in August when he could examine fresh samples. *Heliozoa. Acanthocystis chaetophora* (Schrank) (*A. tuffacea* Archer)

98

Table 24

Rotifera recorded by Wailes (1939) and by Galliford (1947 and 1949) in
Windermere (Wi), Esthwaite (Es) and Blelham (Bl)

Microcodides chlaena (Gosse) Wi
Euchlanis meneta Myers Wi
Keratella cochlearis (Gosse) Wi, Es, Bl
   var. *robusta* Lauterborn Wi
   var. *hispida* Lauterborn Wi
   var. *irregularis f. angulifera* Laut. Wi
   quadrata (Müller) Wi, Es, Bl
   var. *curvicornis* (Ehrb.) Bl
   serrulata (Ehrb.) Bl
Notholca striata (Müller) Wi, Es, Bl
Argonotholca foliacea (Ehrb.) Wi, Es
Kellicottia (*Notholca*) *longispina* (Kellicott) Wi, Es, Bl
Colurella adriatica (Ehrb.) Bl
Lepadella patella (Müller) Wi
Lecane mira (Murray) Wi
Cephalodela exigua (Gosse) Wi
Trichocerca capucina (Wierz. u. Zach.) Wi, Es, Bl
   similis (Wierz.) Wi, Es, Bl
   stylata Gosse Wi
   rousseleti Voigt Wi
   porcellus (Gosse) Wi, Es, Bl
   weberi (Jennings) Wi
Gastropus stylifer (Imhof) Wi, Es, Bl
   hyptopus (Ehrb.) Wi
Chromogaster ovalis (Bergendal) Wi, Es
   testudo (Laut.) Wi, Es, Bl
Asplanchna priodonta (Gosse) Wi, Es, Bl
Polyarthra trigla Ehrb. Wi, Es, Bl
Anarthra aptera (Hood) Es, Bl
Synchaeta grandis (Zach.) Wi, Bl
   kitina (Rouss.) Wi, Es
   oblonga Ehrb. Es
   tremula (Müller) Wi
   pectinata Ehrb. Wi, Es, Bl
Ploesma hudsoni (Imhof) Wi, Es, Bl
Testudinella incisa (Ternetz) Bl
Filinia longiseta (Ehrb.) Wi, Es, Bl
Pedalia mira (Hudson) Wi
Conochiloides dossuarius (Hudson) Es, Bl

Table 24—cont.

Conochilus hippocrepis (Schrank) Wi
  unicornis (Rousselet) Wi, Es
Collotheca pelagica (Rousselet) Wi, Es, Bl
  mutabilis (Hudson) Wi, Es
  libera (Zach.) Wi, Es

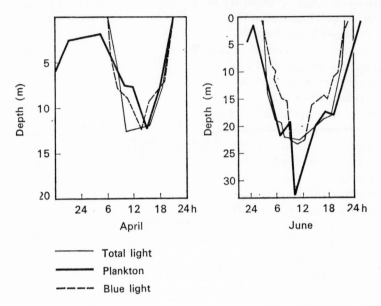

42. Daily migration of *Cyclops strenuus* and the penetration of light in Windermere. The thick continuous line was obtained by plotting at hourly intervals the depth at which most specimens were caught; the thin continuous line shows at hourly intervals the depth to which total light at an intensity of 32,800 erg/cm²/sec in April, and 108,000 erg/cm²/sec in June, penetrated. The thick broken line shows blue light treated in the same way, the intensity taken being 9600 erg/cm²/sec in April, and 305 erg/cm² sec in June (from Ullyott, P. (1939), *Int. Rev. Hydrobiol.* **38**).

appears in June, is numerous by the end of that month and disappears by the end of July. Most specimens are var. *simplex* Wailes.

A. spinifera Greeff is rare.

Rhaphidophrys elegans Hertwig and Lesser occurred in small numbers in August and September.

Elaeorhanis oculea (Archer) (*E. cincta* Greeff) and *Rhaphidiocystis lemani* Penard had been recorded by other workers.

Ciliata. Balanophrya (Holophrya) mamillata Kahl was not infrequent in centrifuged samples.

Tintinnopsis (Tintinnidium) wrayi Wailes was rare in centrifuged samples.

Vorticella sphaerica d'Udeken was usually attached to *Tabellaria*.

V. similis Stokes was usually attached to *Anabaena* cells, occasionally to copepods.

Strombidium viride Stein f. *pelagica* Kahl was found occasionally in centrifuged samples.

Colebrook (1960) studied the effect of water movements on the distribution of plankton. Ullyott (1939) studied vertical migrations of *Cyclops strenuus*, and found close agreement between the depth at which the animals were thickest and the depth to which blue light was penetrating (fig. 42).

Mortimer (1939) attempted what he later (Mortimer 1959) described as 'a preliminary, naïve attempt' to calculate production by the plankton. His figure (reproduced as fig. 43 here) shows actually the standing crop on

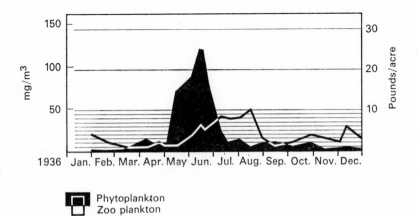

43. Plankton production (dry weight) in Windermere, assuming uniform distribution (Mortimer, C. H. (1939), *Proc. Assoc. appl. Biol.* **26**).

various dates of the phytoplankton retained by a net with 72 meshes/cm and the zooplankton retained by a net with 24 meshes/cm. Uniform distribution throughout the lake is assumed.

Smyly (1968), using data of the same kind (fig. 41), confirms that Mortimer's figure is of the right shape but does not commit himself to a calculation of production. An accurate assessment of production from such figures requires a knowledge, as shown later for *Salmo trutta* by K. R. Allen in New Zealand, of the number of eggs produced by each species and of the survival at frequent intervals thereafter (Allen 1952). A start on work of this kind has been made by Smyly (1961) who has studied *Cyclops leuckarti* in Esthwaite. Egg-laying starts in May and the females produce batches of 15–36 eggs over a period which may be as long as 40 days. A second generation, which reaches a peak in July, produces 5–10 eggs. The progeny develop to the late copepodid stages and then retire to the bottom of the lake at the time of overturn, ultimately to hibernate in the mud. In winter the species is not to be found in the water, and the copepodids, mostly in stage V, a few in stage IV, reappear in February or early March and complete development. The factors which govern the onset and end of hibernation are still under investigation. It is always difficult to discover how many eggs are produced by females that lay a number of batches, and the figures on which production might be calculated are still not available.

Cyclops albidus, C. leuckarti, C. strenuus, and *C. viridis* are carnivores, feeding on crustacea, chironomid larvae and oligochaetes. Other species, *C. agilis* for example, are herbivorous (Fryer 1957). Dr G. Fryer's extensive work on the methods of feeding of Cladocera and Copepoda lies outside the scope of this book.

CHAPTER 8

Rooted Vegetation

Pearsall described the vegetation of Esthwaite Water in 1917 and discussed the higher plants of all the lakes in 1920, in which paper he wrote: 'While the writer has been in intimate contact with three of the lakes for the last fifteen years, the work on which this account is based was started in 1913, and continued, with considerable interruption, until 1920.' The 'considerable interruption' was war service. Since then, apart from a mainly chemical study by R. D. Misra, who was working for a Ph.D. under Pearsall's supervision, hardly any work has been done on the higher plants. The studies mentioned form the basis of chapters 30 and 32 in Tansley's (1939) *The British Islands and their Vegetation.*

The terms 'rooted' vegetation, or 'higher' plants, include the Characeae and various mosses which, strictly speaking, are neither. Ecologically they form, with the phanerogams, a narrow zone between the line below which there is insufficient light and a line above which there is too much wave action; in sheltered places land and aquatic vegetation merge. In productive lakes plants extend down to a depth where the light is about 2 per cent of the intensity at the surface, but in the unproductive lakes they stop short at a line 1–2 m above this. Pearsall and Hewitt (1933) record that the lower limit of rooted plants in Windermere had risen from 6·5 m in 1920 to 4·3 m in 1932. According to Godward (1937) it was at 6 m during the years 1933, 1934 and 1935.

Pearsall not only described the distribution of species but made good progress with the task of explaining it. This was largely a matter of deduction from field observations, though some chemical analysis was done, an aspect of the work which, Pearsall points out, was rendered difficult by the absence of any form of laboratory accommodation.

Species tend to occur in zones along the isobaths. This is demonstrated particularly well in fig. 44 which shows the vegetation along a stretch of the west shore of Windermere between Watbarrow Point and Wray Crag (fig. 15). In order to include a long stretch of shore, it was necessary to make a map so small that the isobaths could not be included without causing confusion. They are shown on fig. 45 which depicts a portion of the area on a larger scale. One factor which varies regularly with depth is light, but Pearsall soon noticed that no species occurs in a constant position in relation to light, and sometimes the usual order of plants is reversed; for example, off the north shore of Ennerdale *Isoetes* is in deeper water than *Nitella*. There is, however, some distinction between the shallow-water and

103

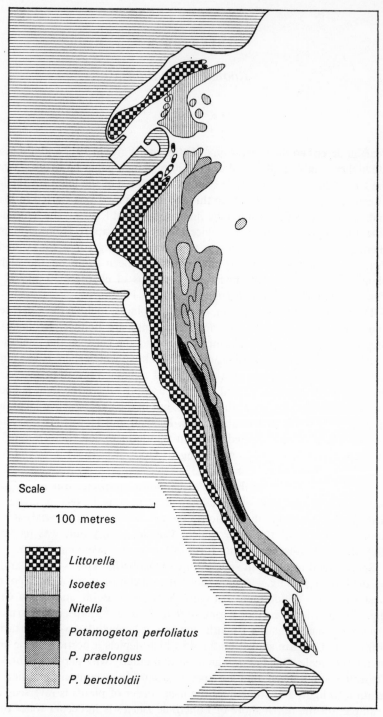

Scale

100 metres

Littorella

Isoetes

Nitella

Potamogeton perfoliatus

P. praelongus

P. berchtoldii

44. Vegetation between Wray Crag and Watbarrow Point on the west shore of Windermere (from an unpublished survey by W. H. Pearsall). *P. berchtoldii* = *P. pusillus*

the deep-water communities. Many constituents of the former are emergent or floating-leaved plants, whose form no doubt restricts them to shallow water, though how it does so has not yet been explained, but *Littorella* and *Lobelia* are not. Their absence from deeper water is probably related to an inability to function efficiently in dim light.

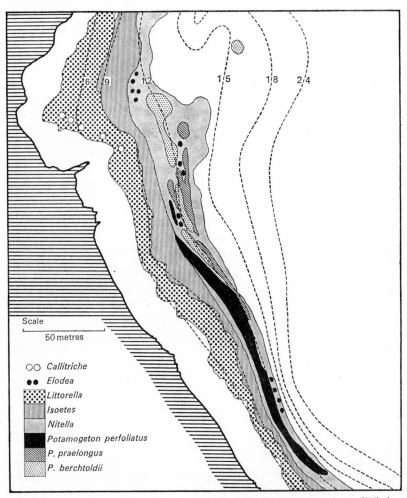

45. Vegetation between Wray Crag and Watbarrow Point, on the west shore of Windermere (an enlarged portion of fig. 44). The level of the lake was 130 feet above sea level at the time of the survey. Isobaths in feet. *P. berchtoldii = P. pusillus*

Another factor which varies with depth is the fineness of the sediment that falls to the bottom. There is, however, no exact relationship between particles of a given size and a particular depth; settling is determined by the amount of wave action and this may be the same in deep water off an exposed shore and in shallow water in a sheltered bay. Pearsall soon observed a correlation between the bottom deposits and the species of plant, and he illustrates it with a series of examples. Two islands in Derwentwater (fig. 46) are without vegetation on the exposed sides, but

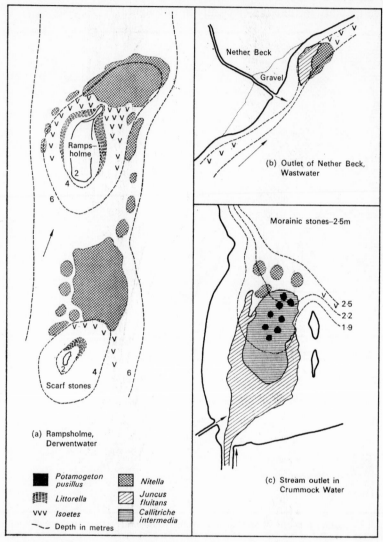

46. Distribution of vegetation in relation to silting effects (Pearsall, W. H. (1920), *J. Ecol.* **8**).

colonized on the sheltered sides, where banks of gravel, with silt in the deeper water, have been deposited. *Isoetes* occurs in the shallower water where the rate of silting is slow and sparse silt covers coarser deposits. *Nitella* grows beyond in deep silt, which is obviously being deposited more rapidly. Considerable areas in all the lakes are occupied by these two species, *Isoetes* being found in those areas where little settles, and *Nitella* in deeper water where the bottom is covered with silt.

Three species occur in Wastwater on the side of a delta which is sheltered from the prevailing wind, all in the same depth of water. *Juncus fluitans*

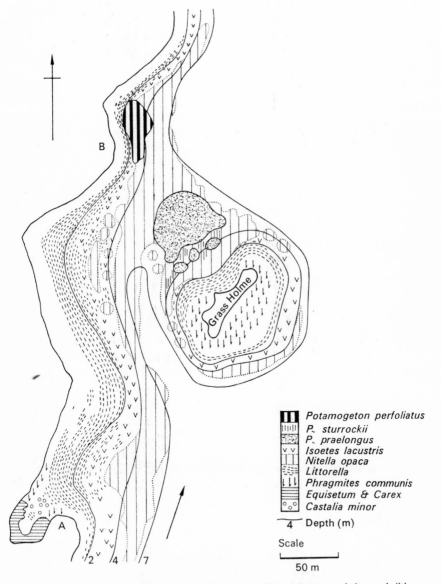

Legend:

- **Ⅲ** *Potamogeton perfoliatus*
- *P. sturrockii*
- *P. praelongus*
- *Isoetes lacustris*
- *Nitella opaca*
- *Littorella*
- *Phragmites communis*
- *Equisetum & Carex*
- *Castalia minor*

4 Depth (m)

Scale

50 m

47. Grass Holme, Windermere, showing the relation of *Potamogeton* to shelter and silting, also a closed bay with *Castalia minor* and *Equisetum-Carex* reed swamp (Pearsall, W. H. (1920), *J. Ecol.* **8**). *P. sturrockii* = *P. pusillus*

is nearest the stream which is building up the delta, then comes *Nitella* and beyond it *Isoetes* (fig. 46). *Juncus fluitans* is obviously growing where sediments are accumulating more rapidly than in the area occupied by *Nitella*. In contrast, under the lee of Grass Holm, an island in Windermere, species of *Potamogeton* are growing where sediments are being deposited and *Juncus fluitans* does not occur anywhere in the vicinity (fig. 47). There should be a zone of it if the requirements of the *Potamogeton* species were no more than a fall of silt more copious than that which creates the substratum on which *Juncus fluitans* flourishes. Its absence suggests a qualitative difference as well. Light on this is shed by the species found at the mouth of the main inflow into Crummock Water (fig. 46). First comes *Juncus fluitans*, then *Callitriche intermedia*, and in it, but in that part which is farthest from the river mouth, *Potamogeton pusillus*. The amount of organic matter in the sediments on which these plants are growing is similar, which suggests that the rate of silt deposition is uniform. There will, however, be a grading of the silts as they enter the lake, the coarsest coming down first and the finest travelling furthest. *Potamogeton pusillus*, it appears from this, requires not only a fairly rapid deposition of silt, but deposition of relatively fine silt.

Fine silts are to be expected predominantly in the well weathered drainage areas in which the productive lakes lie, coarse silts in the rocky drainage areas of the unproductive lakes.

DEEP-WATER COMMUNITIES

Pearsall recognizes twelve different deep-water communities.

1. The *Isoetes lacustris* consocies occurs on stones thinly covered with silt and on boulder clay. It may also occur on sand or on thin peat. The content of organic matter in the soil ranges from 2 per cent to 70 per cent. It appears to be unable to alter its rooting level, and is smothered if sediment deposition is not extremely slow. It covers great areas of the unproductive lakes and, in productive ones, occurs on the face of the wave-cut platform where wave action is sufficient to move silt but not stones. It is usually the only species and the individual plants are placed well apart.

On a highly organic substratum *Utricularia* and mosses are associated with it, on an inorganic substratum with some silting the other plants in the following list:

Isoetes lacustris L.	d	*Utricularia neglecta* Lehm.	l
Nitella opaca Ag.	la	*U. ochroleuca* Hartm.	l
N. flexilis Ag.	l	*Fontinalis antipyretica* L.	l
Chara fragilis Desv.	r	*Eurhynchium rusciforme* Milde	r
Potamogeton perfoliatus L.	l	*E. praelongum* Hobk.	r
Sparganium minimum Fr.	r	*Climacium dendroides* W. and M.	r
Myriophyllum spicatum L.	la	*Hypnum cuspidatum* L.	r
Ranunculus spp.	r	*Mnium cuspidatum* Hedw.	vr

2. The *Nitella* associes can establish itself only on a soft substratum, whose organic content ranges from 8 to 35 per cent. The sediment, as already noted, is generally relatively coarse and the rate of accumulation is slow. This is another community found widely in lakes of all types. The composition is:

Nitella flexilis Ag.	ld	*Potamogeton pusillus* L. subsp.	
N. opaca Ag.	d	*lacustris* subsp. nov.[1]	o
Chara fragilis Desv.	o	*P. crispus* L.	r
C. fragilis sub sp. *delicatula*		*Elodea canadensis* Michx	l
Braun.	la	*Myriophyllum spicatum* L.	f
Isoetes lacustris L.	l	*M. alterniflorum* D.C.	l
Potamogeton praelongus Wulf.	l	*Callitriche intermedia* Hoffm.	r
P. perfoliatus L.	r	*Eurhynchium rusciforme* Milde	r
P. pusillus L.	r		

3. The *Juncus fluitans* consocies occurs in the more unproductive lakes where coarse silts are accumulating rapidly. The species composing the community are:

Juncus bulbosus L. f. *fluitans*		*Isoetes lacustris* L.	f
(var. Lam.)	d	*Potamogeton pusillus* sub sp.	
Callitriche intermedia Hoffm.	la	*lacustris*	l
Myriophyllum spicatum L.	f	*P. polygonifolius* Pourr. var.	
M. alterniflorum D.C.	r	*pseudo fluitans* Syme	r
Nitella opaca Ag.	f	*Utricularia intermedia* Hayne	l
N. flexilis Ag.	l	*Fontinalis antipyretica* L.	l
Chara fragilis Desv.	o		

It also occurs on highly organic soils (49 per cent to 92 per cent organic matter), which share with the other type of substratum the characteristic of a poor supply of bases. Associated with it on organic substrata are *Lobelia* and *Potamogeton natans*. It is one of the few species found in both shallow- and deep-water communities.

4, 5, 6. The next three communities are dominated by species of *Potamogeton* and their composition is shown in table 25. They occur in the more productive lakes where fine silt is accumulating. The *P. pusillus* community is widespread but in general occurs on coarser silts than the others. *P. praelongus* occurs in deeper water than *P. perfoliatus* on finer silts where accumulation is rapid. Figs. 44 and 45 show a typical stretch of the west shore of Windermere. It is a slightly indented bay bounded by two rocky headlands, near which the shore shelves steeply. Near the middle the gradient is less and here the community occurs.

7. *P. alpinus* occurs on soils with more organic matter.

8. *Sparganium minimum* also occurs on organic soils and is frequently associated with *P. alpinus* and *P. obtusifolius*. Often it is the only species and the plants are well spaced, suggesting unfavourable conditions. Pearsall suggests the possibility that successive communities growing and

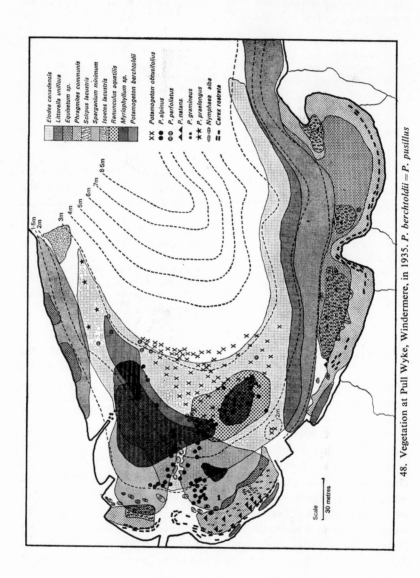

48. Vegetation at Pull Wyke, Windermere, in 1935. *P. berchtoldii = P. pusillus*

dying on the soil have made it toxic. Water-lilies are nearly always colonizing such places.

9. *Fontinalis antipyretica* is often found near the mouths of streams on an organic substratum together with other species characteristic of those conditions. The community, however, is also found colonizing bare rocks rather sparsely.

Fig. 48 shows the vegetation in Pull Wyke in 1935. It is one of the most protected bays in Windermere (fig. 16) and supports the communities typical of an organic substratum. A similar association, with more *Fontinalis*, is found in the North Bay of Esthwaite, which is figured in Tansley (1939). Here much of the organic matter is formed by dead leaves brought down by Black Beck, the main inflow. The stream running into Pull Wyke is much smaller but there is an accumulation of dead leaves, many of them probably blown in.

The last three communities are confined or almost confined to Esthwaite where silting with fine silt is at its most rapid.

10. *Naias flexilis* grows on fine semi-liquid muds with an organic content of only 5 per cent to 10 per cent. It appears to be the first colonist and takes the place occupied elsewhere by *Nitella*.

11. The linear-leaved associes, like the preceding, is confined to Esthwaite. The soil contains a little more organic matter (10 per cent to 15 per cent) than that on which the *Naias flexilis* consocies is found. It resembles the *P. pusillus* consocies of the other lakes, but *P. panormitanus* Biv. B. and *Callitriche autumnalis* L. are sub-dominant and *Elodea canadensis* frequent.

12. The *Potamogeton obtusifolius* consocies had replaced the linear-leaved associes in several parts of Esthwaite since 1915, and occurred in parts of Windermere (fig. 48). It occurs on more organic substrata than that community, and is associated with species typical of a highly organic substratum such as *Elodea canadensis*, *Sparganium minimum* and *P. alpinus*.

The relationships of these species are summarized in fig. 49. Pearsall's caption has been retained but the word 'succession' in it calls for some comment. In the first place it has been inferred from communities seen at a number of different places and not observed at one. The changes on the left-hand side, that is changes in the amount of silt horizontally and in the degree of fineness vertically, are independent of the plants themselves. Changes on the right-hand side, when accumulation of organic matter has become more important than the amount and quality of the silt, depend much more on the plants, which may be the major contributors of the organic matter. It is possible that all the changes shown in the diagram have taken place at some time or other, though if so, some can have affected no more than a small area. It is obvious that many are likely to be rare and that fig. 49 is designed to illustrate the ecological relationships between the species shown and the nature of the substratum rather than actual successions that are likely to take place.

Pearsall recognizes three main successions, characteristic respectively of unproductive lakes, productive lakes and Esthwaite.

Each starts with submerged plants and ends with emergent plants, having passed through a floating-leaved stage. Pearsall's approach is that of a botanist concerned to explain the development of communities, and he is justified in tracing the history of a reed-swamp or a bed of *Carex* back to a stage of floating-leaved vegetation and, before that, submerged aquatics. The limnologist will be aware that this process is confined to small lakes and to bays in larger ones; in other words it is not general. More typical of a lake as a whole is the sequence of plants off a stretch of

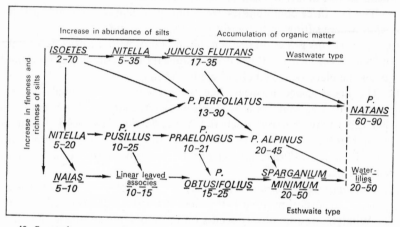

49. Succession among deep-water communities. The figures given indicate the percentage organic content of the soil (Pearsall, W. H. (1920), *J. Ecol.* **8**).

shore such as that shown in fig. 45. Down to a depth of about 2 m the substratum is stony and so pounded by waves that it supports no rooted vegetation. Presumably the stones are broken into small pieces and renewed by destruction of the rock that bounds the lake, but this is an extremely slow process. In deeper water, *Littorella*, able to withstand the wave action, colonizes a stony substratum and below it, on the finer deposits where *Isoetes* grows, there is still enough water-movement to remove much of the silt and the dead remains of plants. The soil, therefore, probably remains unchanged for immensely long periods. Accumulation is negligible but the rain of silt is sufficient to maintain the small quantity of nutrients required. If this state of balance persists as long as seems probable, the communities are nearer an edaphic climax than the first stages of a succession. The deeper parts of a lake are slowly filling up, and in time the whole lake will contain silt up to the depth at which it settles today. What happens next is a matter for conjecture. A small lake would become overgrown with vegetation and the successions described by Pearsall would develop. In a lake as large as Windermere, and with as much water flowing through it, however, the wind stirring it up and floods carrying it away would prevent silt accumulating after a certain depth had been reached, probably one near that at which *Isoetes* ceases today. This cycle of erosion and regeneration might continue almost indefinitely. The erosive forces would be favoured because, once silt had extended upwards to a depth where it came within reach of the waves, the water would be so turbid that the lower limit of vegetation would be reduced. There might indeed be sufficient light only at a depth where the bottom was eroded by the waves too often for plants to be able to establish themselves.

Short-term events in the deep water are open to almost as much speculation as long-term events. Below *Isoetes* there is accumulation of material but whether silt and the organic matter derived from the plants themselves, and from other sources, balance to maintain a soil of unchanging composition is not known. If there is accumulation of organic matter, is there a succession of species and where does it end? In three or four metres of water it certainly does not go to water-lilies, as indicated in fig. 49. Possibly a sterile phase intervenes, micro-organisms decompose the organic matter and the succession starts again at an earlier stage. Only observation can show whether there is a cycle of this kind and, if so, how long it takes.

Actual successions then are to be observed in bays, which are infrequent features of the shorelines of most lakes. Of the deep-water communities off the more usual type of shore it may be said now, by way of summary, that, in the primitive lakes little will be found except *Isoetes* and *Nitella*. Such silts as are brought in are mostly coarse and therefore not carried far beyond the delta of the stream that bears them. In more productive lakes will be found the communities of big *Potamogetons* included in table 25. On the finest silts, largely confined to Esthwaite, are the communities of *Naias flexilis*, the linear-leaved associes and *P. obtusifolius*.

As the Wastwater delta shown in fig. 46 grows, the prevailing current may deflect it to form a spit. If this provides sufficient shelter, organic matter will accumulate and there will be little decomposition owing to the scarcity of silt. *Juncus fluitans* will colonize this highly organic substrate and then give way to *Potamogeton natans*. This is the typical succession of the unproductive lakes.

In bays in Windermere the *Potamogeton alpinus* community is found on soils rich in organic matter and this appears to be followed by a sterile soil on which little but sparse *Sparganium minimum* grows. The floating-leaf stage is water lilies.

Esthwaite is similar except that the *P. obtusifolius* community precedes *Sparganium*.

The change from submerged to floating-leaved plants appears to be

I   113

Table 25
Composition of Potamogeton communities

	4	5		6
	P.	*P.*	*P.*	*P.*
	pusillus	*perfoliatus*	*praelongus*	*alpinus*
% organic matter	12–26	13–29	10–26	20–44
P. pusillus subsp. *lacustris*	d	–	o	–
P. pusillus L.	o	1	r	–
P. obtusifolius M. & K.	r	–	–	l
P. crispus var *serratus* Huds.	vr	–	–	–
P. praelongus Wulf.		o	d	–
P. perfoliatus L.		d	l	–
P. alpinus Balb.	–	r	r	–
P. angustifolius Presl.	–	l	–	–
P. lucens L.	–	–	r	–
Elodea canadensis Michx.	lf	l	l	f
Callitriche intermedia Hoffm.	o	–	–	–
Myriophyllum spicatum L.	f	o	o	f
M. verticillatum L.	r	–	–	l
M. alterniflorum D.C.	r	–	–	–
Nitella opaca Ag.	f	o	f	–
N. flexilis Ag.	l	–	–	–
Chara fragilis Desv.	–	r	–	–
Isoetes lacustris L.	l	l	–	–
Sparganium minimum Fr.	–	–	–	o
Nymphaea lutea L.	–	–	–	o

d = dominant, sd = subdominant, f = frequent, o = occasional, r = rare, vr = very rare, l = local.

irreversible but changes earlier in the succession can take a direction opposite to the usual one if there is an appropriate alteration in the substratum.

SHALLOW-WATER COMMUNITIES
The commonest species is *Littorella*, of which some mention has been made already. Along a typical stretch of lake, such as is shown in fig. 45, it is the topmost rooted plant, growing on an inorganic substratum, which its reproduction by means of stolons enables it to stabilize. Where the hinterland is moraine, it extends into shallower water, and in sheltered

places it is continuous up to a line which is uncovered when the lake level is low. Where the formation of spits provides enough shelter to allow fine sediments to settle other plants come in and Pearsall recognizes:

1. *Potamogeton* associes   on soil with 5 to 40% organic matter.
2. *Myriophyllum* consocies „ „ „   20 to 65% „   „
3. *Juncus fluitans* socies „ „ „   60 to 96% „   „

The *Potamogeton* associes may be composed of one of the following species:

P. *gramineus* L.	P. *angustifolius* Presl.
P. *gramineus* var. *longipedunculata* (Merat.)	P. *perfoliatus* L.
P. *nitens* var. *subgramineus* Raunk.	P. *alpinus* Balb.

This associes is found occasionally in productive lakes.

The *Myriophyllum* is common where dead leaves accumulate and often forms a line along the base of the wave-cut terrace. *M. spicatum* L. is the most frequent species except in Esthwaite where *M. alterniflorum* D.C. takes its place. *M. verticillatum* occurs rarely and *Ranunculus peltatus* Schr. var. *truncatus* Koch may be locally abundant.

The *Juncus fluitans* socies occurs on peaty soils, on which *Lobelia* is commoner than *Littorella*.

Floating-leaved communities develop in sheltered bays, and which of four species occurs appears to be determined largely by the amount of organic matter in the soil, thus:

Castalia alba	10 to 35%
Nymphaea lutea	24 to 55%
Castalia minor	10 to 75%
Potamogeton natans	60 to 92%

However, exposure plays some part and the more inorganic substrata tend to be colonized by *Castalia minor* and *Nymphaea intermedia* where there is some wave action.

No floating-leaved plants occur in many of the lakes. Derwentwater is one where the coastline is irregular, and in this lake, where only moderate amounts of silt occur, the species found is *P. natans* except occasionally near the mouths of streams. *Castalia alba* fringes the reed-swamp at the mouth of Black Beck, the main inflow of Esthwaite, where silt is fine and copious in amount (Tansley 1939 figs. 114 and 115).

The emergent species show a similar relationship to organic matter:

Typha latifolia consocies	20 to 50%
Scirpus—Phragmites associes	20 to 65%
Equisetum limosum	40 to 80%
Carex spp. associes	80 to 95%

Phragmites communis, and *Scirpus lacustris*, which generally grows out-side it in deeper water, are the usual emergents in Windermere and Esthwaite. *Typha latifolia* occurs as well near the mouth of Black Beck in Esthwaite.

The development of reeds impedes the flow of water and diminishes the supply of silt. Consequently the plant remains accumulate and the soil becomes more organic. Pull Wyke (fig. 48) is a typical bay with a small inflow and here there is an extensive growth of *Equisetum* and a regular appearance of *Carex rostrata* on the inside of the *Phragmites* beds. This species is more tolerant of an organic soil than *Phragmites* and gradually replaces it. Its presence suggests evolution in the direction of bog. In the North Bay of Esthwaite, where there is a larger stream well laden with silt, *Phragmites* persists longer and, when Pearsall surveyed the area in 1914, much of it was above the water level and growing in association with *Salix atrocinerea*. In other words, under these conditions evolution was in the direction of fen. *Carices* occurred only in the margin of the reed-swamp. He repeated the survey in 1929, and his account and map appeared in chapter 30 and fig. 115 of Tansley's (1939) *The British Islands and Their Vegetation*. During the fifteen years the reed-swamp had advanced between 50 and 100 feet (17–34 m) into the lake, but on the landward side much of the *Phragmites* had disappeared. In the area furthest from the inflow *Carex inflata* had increased considerably.

The floating-leaved and emergent communities cannot be used to indicate the position of a lake in the series, because their presence depends either on small size or on the occurrence of bays. In large lakes with regular shore-lines they do not occur. The deep-water communities are more useful for this purpose and Pearsall (1921) discusses them from this point of view, and summarizes his findings in a table, reproduced here in slightly modified form (table 26). In the first four lakes, the most un-productive, *Isoetes* and *Nitella*, are by far the commonest plants, and there is little else apart from some *Juncus fluitans* and *Myriophyllum*, mainly *spicatum*. Pearsall notes that considerable areas of suitable depth and apparently suitable substratum are without plants. In Haweswater there is an un-usually high proportion of *Nitella*, which is growing on extensive areas of fine sand laid down during a period of intense erosion, probably in the final stages of the Ice Age. In Bassenthwaite, on the other hand, there is little *Nitella* and the percentage of *Potamogeton* is not as high as might be expected in a lake of its type. This is due to the stained water (table 13, p. 56), which absorbs light more rapidly than that of the other lakes, with the result that insufficient light for photosynthesis penetrates to the depths at which these species would otherwise be growing. There is more *Pota-mogeton* than might be expected in Coniston and Ullswater, and this is attributed to silt brought in with mine washings. On the whole, therefore, a study of the rooted vegetation yields a less convincing serial arrangement

Table 26

Percentages per lake of plants growing below 2 m. (Pearsall, 1921. *Proc. R. Soc. B.* **92**)

	Wa	En	Bu	Cr	Ha	De	Ba	Co	Wi	Ul	Es
Isoetes lacustris L.	49	35	40	48	5	31	42	34	9	34	2
Nitella opaca Ag. / *N. flexilis* Ag.	36	46	40	26	65	42	3	9	40	15	26
Chara fragilis Desv.	–	2	–	–	6	1	–	–	3	–	2
Total	85	83	80	74	76	74	45	43	52	49	30
Juncus bulbosus L. f. *fluitans* Lam.	6	5	7	8	1	7	2	2	–	–	–
Callitriche intermedia Hoffm.	2	1·5	3	5	4	1	21	7	0·5	–	1·5
Potamogeton pusillus L.	0·5	0·5	1	2	2	1	3	15	3	10	6·5
P. perfoliatus Wulf.	–	–	–	–	0·5	4	–	13	11	17	11
P. praelongus Wulf.	–	–	–	–	–	1	–	2	14	8	1
Other *Potamogetons*	–	–	–	–	2·5	–	–	–	1	–	7
Elodea canadensis Michx.	–	–	–	–	–	–	–	–	9	–	10
Naias flexilis R. & S.	–	–	–	–	–	–	–	–	–	–	21
Myriophyllum spp.	6·5	10	9	8	13	2	11	6	4	14	2
Others	–	–	–	3	1	10	18	12	2·5	3	10

117

than some others, because peculiarities of individual lakes modify the effect of the general nature of the lake and its drainage area.

Misra (1938) worked mainly with three types of sediment from Windermere, chosen according to the species of plant growing on it:

1. *Isoetes*—inorganic coarse brown silt;
2. *Potamogeton perfoliatus*—moderately organic brown mud;
3. *Sparganium minimum* and *Potamogeton alpinus*—highly organic brown mud.

Had he worked on a less productive lake he might have added a fourth type, the highly organic peat which supports successively *Lobelia* and *Juncus fluitans*, *Potamogeton natans* and *Carex rostrata*.

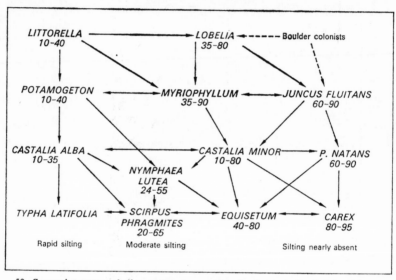

50. Succession among shallow-water communities, the figures indicating the percentage organic content of the soil (Pearsall, W. H. (1920), *J. Ecol.* **8**).

Pearsall had made some chemical analyses, and had found the coarse soils, where *Isoetes* grows, to be poor in nutrients. These, particularly potassium and phosphate, are brought in by silt and the finer the silt the richer the supply. Areas where silting is rapid support a vigorous stand of plants for two reasons, first the nutrients are directly available and secondly something in the silt stimulates decomposition and prevents the formation of a highly organic mud with properties intolerable to some species. For example, where, as off the mouth of Black Beck in Esthwaite, silt falls on the remains of *Phragmites* accumulating at the back of a bed, these decompose to give a black mud. On this *Phragmites* persists after the surface level has risen above normal lake level and eventually gives place to the carr formation of fen. Where there is less silt, as in bays into which no stream or only a small stream flows, the *Phragmites* remains decompose less and gradually from a highly organic soil on which *Phragmites* cannot grow. *Carex rostrata* takes its place.

Misra (1938) concluded that decomposition in these soils is anaerobic, for he found sulphide in 15 out of 17 samples, the two exceptions having a low organic content (3·9 per cent and 5·8 per cent of dry weight). The production of CH_4, commonly observed in lake muds, is further evidence.

In the inorganic muds (type 1) and those highly organic (type 3) the C/N was similar, but, in the moderately organic muds (type 2), there was relatively more nitrogen than carbon. Misra suggests that in these muds CO_2 or CH_4 produced by decomposition is lost, whereas NH_3 is absorbed by complexes. Mackereth (1966) has recently cast doubt on the reliability of Misra's figures for carbon and nitrogen.

Misra tried to extract ions from muds in two ways. First he allowed samples to settle in distilled water, excluded oxygen by means of a layer of paraffin, and extracted samples for analysis at intervals. In the second he leached mud with a normal solution of ammonium chloride. He believed that the processes are similar in both experiments; in the second there is displacement of other ions by NH_3 which has been added, in the first displacement by NH_3 produced by decomposition. The first probably yields the more satisfactory results. Oxygen is absorbed rapidly from the water and is reduced to about 0·5 cc/l after one week. Phosphate comes out of soils 1, 2. and 3. in roughly the proportions 1 : 2 : 4, most during the first week. More CO_2 and more NH_3 came out of type 3. muds, but most HCO_3 came from muds of type 2., presumably because they had most exchangeable bases. A calculation of the base saturation made in the leaching experiment showed a ratio in soils 1, 2, and 3 of 2 : 6 : 1. Hydrogen ions were much more abundant in type 3 soils than in type 2. At the end of seven weeks in the first experiment the following amounts of four metals were found in the water above the muds:

	Fe	Al	Mn	Ca
1.	0·2	58	6	39
2.	5	68	4	47
3.	32	78	0·8	2

Misra's conclusion is that 'the high fertility of the *P. perfoliatus* type of mud (type 2) is associated both with a higher proportion of available ammonia (probably a high rate of production) and also with a higher degree of saturation with bases. On the other hand, the sterility of the most organic muds may be ascribed to the low proportions of available calcium and the high proportions of iron under "natural" leaching conditions as well as to the high exchangeable hydrogen.'

119

Table 27

Composition of plants from Windermere. (Misra, 1938. *J. Ecol.* **26**)

Soil no.	Littorella 2	Isoetes 7	Potamogeton per-foliatus 24	Nitella opaca 24	Sparganium minimum		Potamogeton alpinus 35
					35	85	
Ash	14·66	16·82	17·59	23·64	26·92	21·65	27·49
CaO	1·67	0·74	2·01	1·73	2·78	1·23	1·49
MgO	0·47	0·46	0·73	0·36	0·41	0·47	0·15
Fe_2O_3	0·29	0·39	0·46	0·85	4·16	6·25	6·20
Mn_3O_4	0·05	Tr.	0·02	0·04	0·04	0·19	–
Al_2O_3	1·37	2·30	1·57	1·79	1·09	2·00	1·72
Total N	2·80	2·76	6·25	7·09	4·97	3·28	4·74

Table 27, Misra's analysis of six species, shows a certain correspondence between the composition of the plants and composition of the substratum on which they grow: *Isoetes* contains less calcium than any other species; *P. perfoliatus* and *Nitella* are notably rich in nitrogen; in the two species from the highly organic soil there is more iron than in the rest. Further experiments indicated that at first an increase in the amount of organic matter in the soil led to an increase in the amount of both iron and calcium in *Potamogeton perfoliatus,* but after a further rise the content of iron continued to go up while that of calcium went down.

Experiments in which three species were grown in each of the three types of mud produced the following result:

Jars	*P. perfoliatus*	*P. alpinus*	*Sparganium minimum*
1	Grew well for 20 days and then died	Grew well but narrower and greener leaves produced than found in the lake	Slow growth
2	Perfect growth until the end of the exp.	Poor growth but survived till the end of the exp.	Growth poor
3	Died within a week	Grew well with normal type leaves	Grew vigorously, many, plants came out of the rhizomes
Duration of exps.	4.iv.35 to 24.v.35	24. iv. 35 to 24. v. 35	12. v. 35 to 24. v. 35

This suggests that *P. perfoliatus* and *S. minimum* are confined to soils of a particular type by peculiar requirements. Why *P. alpinus* should grow well on an inorganic substratum in the experiment but not in nature is inexplicable. That this species did not do well on soils of type 2 probably means that it is not competition which keeps it from such places in the lake.

There can be little doubt that the nature of the soil is a factor of great importance in the ecology of rooted aquatic plants. As a result of Pearsall's pioneer surveys, it is possible to describe the various communities and the factors which influence their occurrence. Not much is known about how rapidly they change. Some, it has been suggested above, last for a very long time. Others are changing at a rate observable in a human lifetime, as Pearsall's surveys of the North Bay of Esthwaite and occasional remarks in his papers indicate. Only regular observations, possibly made much more frequently than were Pearsall's, on a few places will throw more light on the speed and nature of these changes. They should also throw light on how far competition determines the nature of a community, an unexplored field at present. *Potamogetons* die down in the autumn and survive the winter in the form of rosettes. It would seem likely that the successful species is the one which starts growth first or which can make up for a late start by growing faster than its rivals. It will be recalled that competition of this nature is important in the algae. This group reminds us too that inorganic substances, that is those which receive most attention from the analyst, may be less important than organic substances whose identity remains unknown. It is hard not to believe that this is true of the mud also. Misra's conclusion, quoted verbatim above, may well prove to be basically true but future work will probably add many complications.

CHAPTER 9

The Bottom Fauna: Studies of Groups

The Wests produced a list of the plankton algae of the lakes in 1909; some of the comparable work on animals is appearing for the first time in this chapter. The reason why animal ecology has lagged so far behind is largely taxonomic. The systematics of most groups were taken to a reasonable stage of completion in Victorian times, or at least before the First World War, but the work extended to the adults only, and did not include the nymphs or larvae, the stage in which most of the life span is often passed and in which the species are encountered by the freshwater biologist. In waters of the type under discussion, the nymphs of Plecoptera and Ephemeroptera and the larvae of Trichoptera and Diptera, especially Chironomidae, are taken often and in large numbers. Work on these started long ago, but the first complete key to any one of them was that which Hynes (1958) published to the Plecoptera in the *Scientific Publications of the Freshwater Biological Association* (No. 17). A similar key to the nymphs of Ephemeroptera (No. 20) appeared three years later (Macan 1961a). Much remains to be done on the Trichoptera and Diptera. Keys to all the species in some families of Trichoptera have been produced by Mackereth (1954, 1956) (Rhyacophilidae) and Edington (1964) (Poly-centropidae), and many single species have been described, mainly by N. E. Hickin (1967), but in the absence of descriptions of closely related species, these cannot be used by the ecologist. Apart from those in the families named above, his only method of identifying the species in any particular biotope is to trap the adults as they emerge.

Investigators starting work on a group which has been studied by taxonomists have often found that, by the time they had included new records, rejected incorrect ones, and brought the existing keys up to date, there was material for a new publication. Such, for example, are the *Freshwater Biological Association's Scientific Publications* on triclads (No. 23, Reynoldson 1967), oligochaetes (No. 22, Brinkhurst 1963) and leeches (No. 14, Mann 1965, 2nd ed.).

The presence of a species depends on whether conditions are suitable for it, and whether it has ever reached the place in question. During the advance of the Ice Age the fauna of Europe was forced to migrate southwards, and the number of species that remained in Britain must have been small, for the ice sheets extended as far south as the valley of the Thames. When the European fauna was returning northwards as the Ice Age receded, Britain was a small projection from one side of Europe and later, some

8000 years ago, it became isolated. The result of the eccentric position and the isolation is that a number of species never succeeded in reaching it. There are, however, few local contributions to the history of the European fauna, a topic on which Thienemann (1950) has dwelt ably and at length, and the subject will not be pursued further here except to record some recent immigrations. An account of the discovery and spread of a new-comer must obviously precede an account of the communities which they have joined. After that the ideal chapter would be devoted to the different biotopes, stony substratum, sandy substratum, submerged vegetation and reed beds, and the biocoenoses inhabiting each. The present one falls short of the ideal because only the first of these biocoenoses has been investigated with any thoroughness. On the other hand, some groups of animals have been studied in detail, and as members of them may occur in all the biotopes mentioned, it is still necessary to treat this work separately.

NEWCOMERS

Planorbis carinatus is not among the species which Boycott (1936, p. 151) records in Windermere, though today it is one of the most numerous. The first report of it was that of Moon (1936). It is impossible to be certain that the earlier collectors, whose records are summarized by Boycott, did not overlook this species but this is unlikely unless it was much less abundant than it is now. It is a species generally confined to water with not less than 20 mg/l Ca and its occurrence in Windermere is, therefore, unexpected. The enrichment of Windermere has, however, probably been the factor which has made colonization by *P. carinatus* possible. It is certainly to be expected that this has enabled new species to come in.

The other two species can be designated newcomers without any doubt, because both are natives of other zoogeographical regions, whose appearance and spread have been well documented. *Crangonyx pseudogracilis* is Nearctic, and *Potamopyrgus jenkinsi* probably Australasian.

Potamopyrgus is also absent from Boycott's list of species in Windermere. It was first noticed in 1936 in Whitecross Bay on the east side of the North Basin (fig. 51), and a survey was made in the following summer. Very large numbers, of the order of 10,000/m², were found on species of *Potamogeton* and other plants in a steep-sided trench cut by a sand and gravel dredger. Numbers decreased rapidly with increasing distance from the edge of the trench, and no specimens were found beyond a line 100 m away. The substratum was sandy and there was a moderately thick growth of *Littorella*. Further north a few specimens were found at three different places and to the south there was a larger colony. The shallow water of the North Basin had been explored at that time, but little work had been done in the South Basin.

Ten years later Whitecross Bay had been altered by the erection beside it of a factory for assembling sea-planes. A slipway and a pier had been

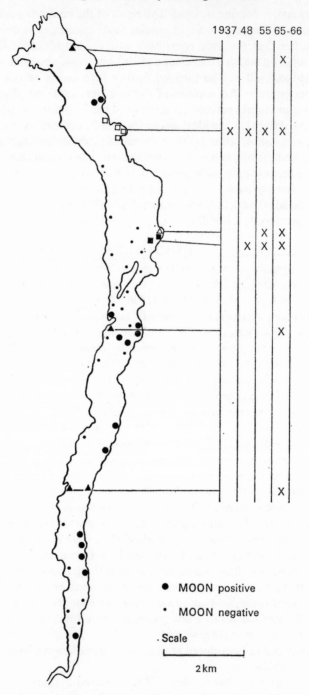

51. Windermere, showing records of *Potamopyrgus jenkinsi*.

built in part of the area where *Potamopyrgus* had been numerous. It was now comparatively scarce. In 1948 it was found in great numbers in a stream flowing into Windermere, and a few specimens occurred at the mouth in the lake. During subsequent years they spread into an adjoining bay. Moon (1956) working at about this time failed to find *Potamopyrgus* in 628 collections at 390 stations in water shallow enough for collecting by hand. He did show that this snail is at its most abundant in water beyond the reach of the collector who has no boat, and extends down to a depth of about 6 m. He found it in 15 out of 44 collections with dredge or grab (fig. 51).

In 1965 and 1966 *Potamopyrgus* was found at five more places during the course of extensive hand-collecting on the stony substratum. At only one, near the outfall of the Bowness–Windermere sewage works, were more than a few specimens encountered. Visits to the places where the snail had been found in earlier years showed that it was still present. Of the five new places three were in the South Basin, which had not been investigated carefully before. The two in the North Basin probably represent an extension of range since 1936 but it would be imprudent not to stress that small numbers of a small animal are easily overlooked. If *Potamopyrgus* is extending its range, it is doing so slowly, and the generally small numbers of this species, of which dense populations are characteristic, indicate that conditions in Windermere are not favourable.

In 1937 W. M. Tattersall wrote that 'some years ago' he had identified as *Crangonyx pseudogracilis* an amphipod that had come out of a tap in Hackney, but had refrained from publishing the record because the emergence of an American species from a London tap seemed improbable. He did not make this occurrence known until Crawford (1937) recorded the species from a water-treatment plant of the Metropolitan Water Board. Hynes (1955) published a map showing its occurrence almost exclusively in counties lying within the canal system of England. It is impossible to separate on such a map the actual spread of a species and the spread of interest aroused by a first record and a taxonomic description, but Hynes observed quick colonization of an area whose fauna he knew well, and the present records indicate that the species is spreading rapidly.

In 1960 a class of students, some of them pupils of Dr Hynes, recorded *Crangonyx* in the South Basin of Windermere, and in the summer of 1961 Mrs E. M. Garland undertook to search the whole lake for it. She found it at almost every station in the South Basin and among the islands in the middle, but not at any of twenty-three stations in the rest of the North Basin. By 1965 it was everywhere in shallow water in the lake, and was found also in Derwentwater, Bassenthwaite and Ullswater.

Crangonyx could have reached Kendal by canal, but even if it reached that town before the canal was filled in, it was still faced by an overland journey of 13 km to Windermere, and a longer one to the other lakes. Nothing is known about how land barriers are crossed.

Moon (1968) has recorded the history of *Asellus* in Esthwaite, a lake whose northern reedswamp has been visited in most years since 1942 by the annual Easter class of students. They did not record *Asellus* in 1942, 1947, 1948, 1949 or 1951, but they did find it in 1944, 1945 and 1950. The number of specimens taken is not known, but this irregular occurrence in the collections suggests low numbers, possibly a small population only just managing to maintain itself. In 1955 the supply department of the Freshwater Biological Association found a few specimens in the same place. In 1956 they thought it was more numerous, and in 1957 noticed that it had colonized the whole of the north bay. During the next few years *Asellus* spread rapidly, and the spread could be mapped accurately as sufficient collections were brought in from all parts of the lake. In 1958 it had extended beyond the north bay, and in 1959 it had colonized about three-quarters of the lake on both sides. In 1961 the only stations at which it was not found were at the extreme end near the outflow, and in 1962 it was found everywhere in shallow water and in deeper water to a depth of 6 m.

The earlier records are, inevitably, less satisfactory than they would have been if the future could have been foreseen, but it is likely that this is an example of a new species that establishes a foothold, maintains it precariously for a period of years, and finally consolidates it before a rapid expansion of range. The advent of *Asellus* is probably associated with the enrichment of Esthwaite, but it is surprising that it did not reach this, the most productive of all the lakes, earlier, for it was in Windermere in 1933 (Moon 1934). It was not recorded before the war in the tarns which drain into Esthwaite from the fells lying to the east, but in 1950 and 1951 it was found in Scale Tarn, the topmost of the series. The large snail *Viviparus fasciatus* was observed in this tarn at the same time and it is likely to have been introduced in about 1939 with a consignment of *Limnaea pereger* purchased to provide food for fish from a hatchery in another part of the country. *Asellus* could have come in at the same time. Both animals have flourished in Scale Tarn but the snail has not spread. *Asellus* was not taken in 1949 or 1951 in Wise Een Tarn, which lies just below Scale Tarn, but it was found there in 1958. Latterly *Asellus* was numerous in the ponds at Wray Mires hatchery which receive water from Wise Een, but how long it had been there is unknown. A short journey down a rapid stream would take *Asellus* into Esthwaite from the hatchery.

STUDIES OF GROUPS

Table 28 shows the occurrence of the species of gasteropod in four different biotopes in Windermere. Adjustments have been made so that there are twelve stations to each biotope, and the figures (simplified from the table in Macan (1950) by the removal of fractions) show the number of stations at which each species was recorded. *Limnaea pereger* was taken

Table 28

Number of stations in Windermere at which the various species of mollusc were recorded. The figures have been adjusted to make a total of 12 stations in each group

	Limnaea pereger (Müll.)	Planorbis carinatus Müll.	P. albus Müll.	Valvata piscinalis (Müll.)	Planorbis contortus (L.)	Physa fontinalis (L.)	Limnaea palustris (Müll.)	Planorbis com-planatus* (L.)	P. spirorbis (L.)	Myxas glutinosa (Müll.)	Potamo-pyrgus jenkinsi Smith
Exposed shore	9	7	7	2	3	1	1	–	–	–	1
Reed-bed on a sandy bottom	8	8	9	4	4	1	1	–	–	–	1
Reed-bed on a muddy bottom	9	8	5	5	2	1	2	1	–	–	–
Submerged vegetation	10	4	6	12	7	7	–	3	1	1	–
Total	36	27	27	23	16	10	4	4	1	1	2

* Referred to as *Planorbis fontanus* Lightfoot by Boycott (1936) who uses the name *Planorbis complanatus* Studer for the species more usually called *Planorbis planorbis* (L.).

at most stations and *Planorbis albus* was also widely distributed. *Planorbis carinatus* and *Limnaea palustris* occurred mainly in shallow water, a fact which becomes evident for the former when total numbers are considered. *Valvata piscinalis*, *Physa fontinalis* and *Planorbis contortus* were found most often in the deeper collections.

During current work *Limnaea pereger* and *Physa fontinalis* have been found on the stony substratum with a frequency and an abundance which suggest that they are regular members of the community inhabiting it. All the other species mentioned have been found on stones, but not often and in small numbers, which has been taken to indicate that they have come in from neighbouring biotopes.

To these seven must be added *Ancylus fluviatilis* and *A. lacustris* to give nine common species in Windermere. Boycott (1936) did not record *Planorbis carinatus*, *Potamopyrgus jenkinsi*, *Myxas glutinosa* or *Planorbis complanatus* in Windermere, but his list does include two, *Planorbis crista (nautileus)* and *Valvata cristata*, which were not found by Macan (1950). Moon (1934) records *P. crista* in his table 9 (p. 24) but not in the list on p. 28. *Valvata cristata* was recorded from Blelham in 1950 and from Esthwaite in 1949. It has not been found in Esthwaite since, although much collecting has been done, which suggests that in spite of periodic introductions it cannot establish itself permanently.

Table 29 shows the occurrence of gasteropods in the other lakes. Loweswater has been left out because it was not included in the original survey (Macan 1950), which is the basis of the table. *Planorbis spirorbis* and *Limnaea truncatula*, both found in both Derwentwater and Bassenthwaite, have been omitted, together with *Planorbis complanatus*, of which a single specimen was taken in Coniston. The only possibly common species omitted is *Ancylus lacustris*, left out because it is not captured during the course of collecting the other species.

These omissions having been made, it can be seen in table 29 that the calcareous Baltic Esrom Lake has a much richer fauna than any Lake District lake, richer indeed than is indicated by the figures quoted here, for they include only the species common enough to be included by Berg (1938) in his tables; in the text he mentions nineteen species. Windermere comes next with ten species. The record of *Myxas glutinosa* in Windermere and in no other lake may be due to the fact that Windermere has been explored most thoroughly, specially those parts with submerged vegetation. The occurrence of *Potamopyrgus jenkinsi* and *Planorbis carinatus* is associated with the favourable conditions in this productive lake but their absence from others that are comparable may be because they have not yet reached them. Boycott (1936, p. 163) writes: 'The larger units of water are liable to contain the more Mollusca if only because the chances of importation have been greater and they contain a greater variety of subhabitats.' Windermere, with two holiday resorts beside it, is more thronged than any

Table 29

Occurrence of Gasteropoda in the English Lakes

	Ancylus fluviatilis	Limnaea pereger	Planorbis contortus	Physa fontinalis	Planorbis albus	Valvata piscinalis	Limnaea palustris	Myxas glutinosa	Potamopyrgus jenkinsi	Planorbis carinatus	P. planorbis	P. crista	Neritina fluviatilis	Valvata cristata	Bithynia tentaculata	B. leachii
Wastwater	X	X														
Ennerdale	X	X														
Buttermere	X	X	X													
Crummock	X	X	X	X												
Derwentwater	X	X	X	X	X	X			X							
Bassenthwaite	X	X	X	X	X	X	X									
Coniston	X	X	X	X	X	X	X									
Ullswater	X	X	X	X	X	X	X									
Windermere	X	X	X	X	X	X	X	X	X	X	X					
Esthwaite	X	X	X	X	X	X	X							X		
(Esrom)		X	X	X	X	X	X			X	X	X	X	X	X	X

other lake except possibly Derwentwater, and is popular with fishermen. Chances of importation are therefore high, particularly in comparison with Esthwaite which is privately owned. The first and last clauses of Boycott's sentence constitute a statement of Thienemann's first law of biocoenotics. There are seven regular species in Esthwaite and in the lakes in the middle of the series, and it is only in the most unproductive lakes that fewer are found.

Ancylus fluviatilis and *Limnaea pereger* are two species noted for tolerance. *Planorbis albus* is the species which, excluding these two, penetrates most extensively into tarns, but the present records suggest that *P. contortus* is more tolerant of adverse conditions within the environment of a lake. *P. albus* was not recorded in Ullswater by Macan (1950) and only a few specimens have been found since. Ullswater's peculiar fauna is referred to again later.

Why some species are less tolerant than others, and whether the less tolerant are excluded from the least productive lakes by difficulties in maintaining the osmotic concentration of the body fluids, by insufficient food, by the absence of some auxiliary factor such as that which *Asterionella* must have, or by something else, only further work will reveal.

Of the lamellibranchs, *Sphaerium corneum* and *S. lacustre, Pisidium cinereum, P. personatum, P. obtusale, P. subtruncatum, P. nitidum, P. milium, P. hibernicum, P. lilljeborgii* and *P. pulchellum* have been found in Windermere, but their distribution in other lakes has not been investigated.

The Corixidae in the lakes have been reported on in a series of papers. Macan (1938) discussed the ecology of the group in all types of water, of which the only lake to receive detailed attention was Windermere. The Corixidae of Blelham were the subject of a paper in 1949 and those of some Danish lakes in 1954(a). Macan (1955) brought together in a short paper the data from all the Lake District lakes. In a second paper in 1954(b), miscellaneous records and records of other workers were analysed and the species were assigned to groups. In 1962 records from stations in Windermere that had been visited more than once were brought together in order to demonstrate that the populations remain reasonably constant in composition from one year to another.

Table 30, taken unaltered from Macan (1938), shows the species taken at twenty-six stations in Windermere, which are arranged in descending order of percentage of organic matter in the soil. This was measured by weighing a sample after it had been dried and again after it had been heated to red heat. The collecting was not quantitative and 'at all stations an attempt was made to collect 20 to 30 corixids within the smallest possible area'. These results are shown graphically in fig. 52. *Micronecta poweri* occurred only where the percentage of organic matter was low. It is the only corixid that does not require a large degree of shelter, and is found on the most exposed shores. It is not, however, found everywhere, and appears to be associated with small patches of shelter such as is provided by a tuft of *Fontinalis* or *Myriophyllum*. *S. dorsalis* was taken at every station, but in greatest numbers where the percentage of organic matter was low. Reed-beds in Windermere are isolated in bays in which the original sandy bottom has often not been covered by plant remains. Where this has happened, the remains are generally confined to the centre of the bed and there is little organic matter in the soil at the edge. There is, therefore, a large area where conditions are optimal for *S. dorsalis*, and on them it can build up a large population. The expansion of this into neighbouring areas, which, in isolation, might not harbour the species, has been put forward as the explanation of the wide distribution of *S. dorsalis* in Windermere. As the amount of organic matter in the soil increases, *S. distincta* and *S. fossarum* become more numerous, and, at the places with the highest values, *S. scotti* is the most abundant species.

A digression must be made at this point to explain to any reader who may consult the original papers referred to why *S. dorsalis* has appeared under three different names. According to the keys available at the beginning of the investigation, the species was *S. striata* Linnaeus, but when specimens were collected in Denmark, it became clear that two

Table 30

Occurrence of Corixidae in Windermere. (Macan, T. T. 1938. *J. Anim. Ecol.* 7)

Locality	Vegetation	Nature of bottom	% o.m.s.	Date	M. poweri	S. dorsalis	S. distincta	S. fossarum	S. scotti	S. falleni	H. sahlbergi
Brathay Bar	*Phragmites*	Gravel	—	17 Aug. 1936	—	15	—	—	—	—	—
Brathay Bay N.	,,	Sand	3·0	7 June 1936	50	11	—	—	—	—	—
Whitecross Bay E.	,,	,,	4·0	6 July 1936	100	28	—	—	—	—	—
Brathay Mouth E.	,,	,,	5·5	12 Mar. 1937	—	19	2	—	—	—	—
Sandy Wyke N.	*Littorella*	,,	6·5	7 June 1936	35	17	1	—	—	—	—
Sea Mew Bay	,,	,,	6·5	10 Aug. 1936	35	3	—	—	—	—	—
Littorella sward M.	*Phragmites*	Washed drift	6·5	6 Dec. 1936	—	17	—	—	—	—	—
Littorella sward S.	*Myriophyllum*		6·5	6 Dec. 1936	—	42	—	—	—	—	—
Pull Wyke Stream	*Phragmites, Equisetum*	Sand	9·3	1 June 1936	—	11	—	—	—	3	—
Ecclerigg Bay	*Myriophyllum*	,,	10·0	4 Dec. 1936	—	18	—	—	—	—	—
Gale Naze W.	*Phragmites, Scirpus*	,,	11·5	15 Oct. 1936	—	17	4	—	—	—	—
Whitecross Bay W.	*Phragmites*	Clay	15·0	6 Dec. 1936	—	70	25	2	—	—	—
Waterhead E.	*Scirpus, Juncus*	Sand	15·5	12 Mar. 1937	—	20	40	2	—	—	—
Whitecross Creek	*Carex*	Mud	19·0	6 Dec. 1936	—	24	3	10	—	—	—
Fisherty Bay	*Myriophyllum*	Debris	22·0	21 Apr. 1936	—	15	17	—	—	—	8
Bee Bay a.	*Phragmites, Carex*	Clay	22·4	2 Dec. 1936	—	4	1	18	—	3	—
Congo Mouth W.	*Phragmites, Scirpus, Carex*	Fibrous	22·5	21 Feb. 1937	—	23	29	12	—	4	—
Congo	*Callitriche*	Mud	30·0	4 Oct. 1935	—	28	13	10	—	—	—
Brathay Bay W.	*Scirpus*	Sand	31·0	15 Oct. 1937	—	30	5	—	—	—	—
Pull Wyke BH.	*Phragmites, Carex*	Fibrous	32·5	24 Nov. 1936	—	3	13	8	—	—	—
Congo Mouth E.	*Scirpus*	Debris	35·5	4 Apr. 1937	—	12	27	3	—	2	—
Pull Wyke Swan	*Phragmites, Equisetum, Carex*	Mud	43·0	21 Feb. 1937	—	8	6	18	—	12	—
Congo Mouth N.	*Phragmites, Scirpus, Equisetum, Carex*	Fibrous	44·5	24 Nov. 1936	—	1	16	15	—	—	—
Union Bay	*Phragmites, Carex, Menyanthes*	Sand and debris	49·0	25 Oct. 1936	—	9	9	—	18	—	—
Bee Bay c.	*Phragmites, Scirpus*	Fibrous	51·0	2 Dec. 1936	—	7	9	10	—	1	—
Gale Naze E.	*Phragmites, Equisetum, Carex*	Mud	52·0	20 Aug. 1936	—	32	3	—	35	—	—

131

species were covered by this name. It was assumed that the Linnean *S. striata* was the Danish one and that the English specimens belonged to a new species. Accordingly they received a new name, *S. lacustris*. This name did not last long, but unfortunately long enough to appear in several ecological papers before being shown to be a synonym of *S. dorsalis*, a name which, given by Leach to British material many years previously, had long been regarded as a synonym of *S. striata*. The generic names in the family have had an even more varied history but, as this is less likely to lead to confusion, there is no need to air the controversy here.

Reedswamp is almost continuous along the west side of Esthwaite and isolated beds are confined to the bays on the east side. It is, therefore, to be expected that *S. dorsalis* would be encountered less often and in smaller numbers in Esthwaite than in Windermere, and that the reverse would hold for *S. distincta* and *S. fossarum* (table 31). *S. falleni* is associated with calcareous conditions (Macan 1954b) and its abundance in Esthwaite is no doubt correlated with the productive nature of that lake. *S. semistriata* is also associated with calcareous conditions. It occurred in Esthwaite at the head of bays, in the sort of place in which, in a less productive lake, *S. scotti* might be expected. Discussion of the remaining species may be deferred until Blelham is considered.

Twenty-six stations were worked in Esthwaite, which happened to be the number on which the Windermere figures were based. There were fifty-six stations in Blelham but in table 31 all the figures for this lake have been halved to make the number of stations in each of the three lakes almost identical. Comparison must be made with caution because, when species are abundant only in one type of place, the total caught will depend on how much time was spent collecting there and how much elsewhere. In all three lakes, however, collecting was extensive and evenly spread as far as could be judged, and the figures are believed to be sufficient for the general comparison made here.

Blelham is a small lake almost completely fringed with reedswamp. At some unknown time in the past, the level has been lowered with the result that *Phragmites* now extends to the edge of the old wave-cut platform. Developing within the reeds is a caricetum in which *Carex elata* is generally the important species, though in places, for example on gravel brought down by streams, it is *C. rostrata*. On the south-east side fen is developing, but opposite, where there is a big bay, there is now bog, probably a stage that followed fen.

S. dorsalis occurs at fewer stations and in smaller numbers than in Esthwaite, a decline that is to be expected. *S. falleni* is more widespread but less numerous. It is a species whose exact habitat has eluded definition (Macan 1954b) but it is probably not a species of thick reeds. Two indicator species are *S. distincta* which is scarcer, though as widespread as in the two other lakes, and *S. scotti*, which was not taken. Only a single *S.*

Table 31

Total number of corixids caught in three lakes. The figures in parenthesis show the number of stations at which each species was caught

Lake	Ref.	Number of Stations	Sigara dorsalis	S. distincta	S. fossarum	S. falleni	S. scotti	Cymatia bonsdorffi	Sigara semistriata	Calli-corixa praeusta	Hesper-ocorixa sahlbergi	H. linnei
Windermere	Macan 1938 table 1	26	484 (26)	223 (18)	108 (11)	25 (6)	53 (2)	*		*	8 (1)	–
Esthwaite	–	26	260 (17)	330 (17)	123 (14)	256 (16)	25 (7)	82 (11)	121 (10)	8 (3)	1 (1)	5 (2)
Blelham	Macan 1949	28	100 (15)	45 (16)	215 (25)	29 (11)		206 (22)	1 (1)	84 (18)	10 (2)	15 (4)

* Recorded on a later occasion.

semistriata was taken. Two other indicator species, of which more will be written later, are *Hesperocorixa linnei* and *H. sahlbergi*, both more common and abundant than in Esthwaite and Windermere.

S. fossarum is the most abundant species and recorded at almost every station, probably because there is an extensive area in which conditions are optimal for it, as there was for *S. dorsalis* in Windermere. Other numerous species are *Callicorixa praeusta* and *Cymatia bonsdorffi*. The former is characteristic of productive places but has not been found abundantly in lakes much larger than Blelham. *Cymatia* is frequently associated both with it and with *S. fossarum*. It is a carnivore and therefore not comparable in requirements with the other corixids.

S. dorsalis, *S. fossarum*, *C. praeusta* and *C. bonsdorffi* are all common in the reedswamp. The same is true at stations off the face of the *Carex elata* at the east end of the lake except that *S. dorsalis* and *C. praeusta* are proportionally more abundant at the expense of the other two. At this end of the lake the bottom shelves gently, and there is a wide reed-bed in which not much debris has accumulated. To this can be attributed the abundance of *S. dorsalis*. At the other end of the lake the reed beds are narrower, the slope of the shore steeper, and the substratum more organic. Here *C. bonsdorffi* is the most numerous species and *S. dorsalis* is scarce. *S. fossarum* and *S. distincta* are more numerous than at the other end of the lake and numbers of *C. praeusta* are about the same. Within the *Carex elata*, *S. fossarum* is the only numerous species.

In these three productive lakes, eight species have been taken in abundance, and it is clear that each has a habitat to which it is narrowly confined, though how or why is not known. Something of the nature of these habitats may be learnt by treating them as terms in a series. In Windermere (table 30 and fig. 52) there is obviously a succession as organic matter increases. Experience in other lakes is that *S. dorsalis* does not persist as long when the percentage of organic matter increases in the soil, and it is found in tarns that *H. castanea* replaces *S. scotti* as vegetation becomes thick. The succession is therefore

$$\genfrac{}{}{0pt}{}{Micronecta}{poweri} \rightarrow S.\ dorsalis \rightarrow \genfrac{}{}{0pt}{}{S.\ fossarum}{S.\ distincta} \rightarrow S.\ scotti \rightarrow H.\ castanea$$

From Esthwaite it may be learnt that, under more productive conditions, *S. falleni* may be an important member of the early stages of the succession. The culminating species appears to be *S. semistriata*, but as this species has not been reported from a similar type of place anywhere else, conclusions about it must await further collecting. In this more productive lake *S. scotti* was nowhere abundant as it was in Windermere. Rather more is to be learnt from the Blelham records. *S. scotti* was not taken and *S. distincta* was less abundant than in the other lakes. *S. fossarum* on the other hand was more common and abundant. This suggests a branching of the succession set out above, with *S. distincta*, *S. scotti* and *H. castanea* along the branch which, with accumulating organic matter, leads to bog, and *S. fossarum* along the branch leading to fen. The final species in this succession is *H. sahlbergi*, found in a pool in Esthwaite North Fen and in a fen beside Windermere. *H. linnei* occurred in all parts of Blelham and was

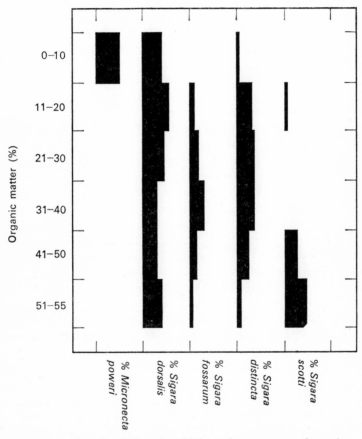

52. Succession of corixid species with increasing percentage of organic matter in the soil of Windermere (Macan, T. T. and Worthington, E. B. (1951), *Life in Lakes and Rivers*).

numerous at one station, but it does not appear to play a prominent part in the succession in the Lake District.

It has already been suggested that *C. praeusta* is a species found in small lakes only. However, not many lakes have been studied, and confirmation of this idea is required.

Esrom Lake is wider than Windermere though not as long. There are thick continuous reed-beds in the bays and isolated ones along the exposed shores. *S. striata* (whose habitat is identical with that of *dorsalis*), *S.*

falleni, C. praeusta, H. linnei and *H. sahlbergi*, all species found in Blelham, were recorded, but only the first two were abundant. *S. striata* was the commonest and most abundant species in the main body of the lake though *S. falleni* occurred at most of the stations. In a bay, where reeds grew densely, the reverse was true. *S. fossarum* occurred in an artificial pond excavated in peat that had obviously developed from reedswamp. In Fure Lake the same species were found, but *H. linnei* was the common species in the innermost parts of the beds. *H. sahlbergi* was found in fen pools. In these productive calcareous lakes there is, therefore, a direct and simple succession:

S. striata→H. linnei→H. sahlbergi. S. falleni is associated with the early stages of the succession. *S. fossarum* was scarce in the lakes but abundant in two places on soil laid down in the lake but no longer in contact with lake water. The absence of *S. distincta* and *S. scotti* confirms their association with places poor in calcium.

A succession: *S. falleni→H. linnei→H. sahlbergi* has been found in Crose Mere and Sweat Mere, two small productive calcareous lakes in Shropshire (Macan 1967).

Comparison of lakes based on the corixid fauna is not always possible, because some lakes do not provide anywhere sufficiently sheltered for these creatures to occur. When they do, comparison must be made with caution because conditions within a reed-bed may reflect some local peculiarity and not the general state of the lake. For example, *S. falleni* was found in Derwentwater in a reed-bed into which flowed a small stream that had passed through a richly cultivated garden. Accordingly, in the summary in table 32, the comparison is based on stations where there is little organic matter in the soil. When this is done, the unproductive lakes are found to be characterized by an abundance of *S. scotti* alongside *S. dorsalis* on gravelly or sandy substrata. Ennerdale, Crummock, Derwentwater, Bassenthwaite and, surprisingly, Loweswater fall into this group. Ullswater, where only *S. dorsalis* was found and Coniston, where there was but a single *S. scotti* to 235 *S. dorsalis*, form an intermediate group. In Windermere, Esthwaite and Blelham *S. scotti* is nearly always absent and *S. falleni* often present in the early stages of the succession.

If well developed reed-beds can be found, comparison can be based on other species as well. *S. fossarum* was fairly or very abundant in Windermere, Esthwaite and Blelham, but was not found in any other lake. *S. distincta*, on the other hand, was taken in all except Ennerdale and Coniston.

It will be noted that the division of the lakes according to their corixids (table 32) does not take them into the same groups that a study of the molluscs does (table 29).

Moon (1957) studied the distribution of *Asellus* in Windermere by collecting with a hand net for two minutes at intervals of 50 m along

Table 32

Occurrence in lakes of the indicator species of Corixidae. (Macan, T. T. 1955. *Verh. int. Ver. Limnol.* **12**)

Oligotrophic ⎯⎯⎯⎯⎯⎯⎯⟶ Eutrophic

	Ennerdale	Crummock	Derwentwater	Bassenthwaite	Loweswater	Coniston	Ullswater	Windermere	Esthwaite	Blelham	Esrom	Fure
S. dorsalis or *striata*	+	+	+	+	+	+	+	+	+	+	+	+
S. scotti	+	+	+	+	+	+			+			
S. falleni								+	+	+	+	+
S. fossarum								+	+	+	+	+

stretches of the shore. He also obtained specimens from deeper water by means of an Ekman Grab or a dredge. The lake could be divided into three parts, roughly equal, except on the east side where the middle part was longer than the other two. In the middle part *Asellus aquaticus* was the species found in the shallow water and here *A. meridianus* was infrequent and never numerous. Below about 2 m, however, *A. meridianus* was the commoner species. Neither was found in water deeper than about 4 m. In collections in shallow water in the northern and southern parts of the lake *A. meridianus* was often the only species found but there were isolated stretches in which *A. aquaticus* was the commoner or the only species.

Moon suggests that *A. meridianus* is the older inhabitant of the lake which the more recent *A. aquaticus* is replacing. The point or points of arrival of the latter would seem to have been near the middle of the lake near Bowness, the main centre of population, and Moon postulates introduction by man.

In the middle part of the lake Moon rarely failed to find specimens but on the western shore of the northern part blank collections were frequent.

Williams (1963) has some evidence that *A. aquaticus* is replacing *A. meridianus*.

Brinkhurst (1964) has collected Oligochaeta in a number of British lakes, but finds that their occurrence is patchy and concludes that more extensive sampling than he was able to carry out (table 33) must be done before the lakes can be compared with each other.

An important gap in faunistic work on the English lakes is the absence of study of groups whose adults are very small. Such animals may possibly contribute significantly to production, rapid reproduction compensating

Table 33

Distribution of Oligochaetes. –absent, +present, + +abundant. (Brinkhurst 1964, *Arch. Hydrobiol* **60**)

Lake	Depth in metres	Limnodrilus hoffmeisteri	Peloscolex ferox	Aulodrilus pluriseta	Tubifex tubifex	Tubifex templetoni	Euilyodrilus hammoniensis
Esthwaite	10	+ +	–	–	+ +	–	+ +
Windermere	4	+ +	+	–	–	–	–
	12	+	+	+ +	–	–	–
	30	+	+	–	–	+	–
Coniston	20	+	–	–	+	–	–
Loweswater	15	+ +	–	–	–	–	–
Bassenthwaite	5	+ +	+	+	–	–	–
Ullswater	25	+ +	–	–	–	–	–
Derwentwater	7	+	+	–	–	–	–
Crummock	–	–	–	–	–	–	–
Buttermere	30	–	+	–	–	–	–
Ennerdale	–	–	–	–	–	–	–
Wastwater	–	–	–	–	–	–	–

for small size, and they must be important as a source of food to larger animals, particularly when these are young. The only contribution so far is that of Dr Marjorie Webb, of Leicester University, who studied the Protozoa in the sediments of Esthwaite (Webb 1961). She recorded more than 120 species, of which about 90 were ciliates, a group that also contributes by far the greatest number of individuals. The study was complicated by two phenomena: first any association is transient, appearing, running through a succession of species and dispersing within a short period, and second, circulation set up by wind mixes the various populations. In calm weather she was able to recognize three groups.

1. Oligosaprobes. Specimens were always few and confined to places where there was plenty of oxygen and little bacterial decomposition. During the period of stratification they were confined to the epilimnion.
2. Mesosaprobes. These abounded at the edge of the area devoid of oxygen, where bacterial decomposition was intense but some oxygen was available.
3. Polysaprobes. These occurred in the hypolimnion where there was no oxygen, but they are facultative anaerobes.

At the edge of the region devoid of oxygen there is a vigorous growth of bacteria which require a little oxygen but which utilize the products of anaerobic bacteria. The main ones are the sulphur bacteria which utilize sulphide, and the iron bacteria which utilize ferrous and manganous compounds. Both include forms which produce a filamentous matrix in which other bacteria and ciliates are numerous. *Spirostomum minus, S. filum* and *Urocentrum turbo* are associated with the growth of the bacteria in laboratory cultures. Later, when decline sets in, *Coleps hirtus, C. octospinus* and *C. incurvus* are the commonest species.

Table 34 shows the commoner species listed by Webb.

Table 34

Ciliata from Esthwaite (Webb 1961). Figures show number of times recorded, and species recorded less than 5 times have been omitted

Group 1		Group 2		Group 3	
Chaenea teres	5	*Coleps hirtus*	42	*Caenomorpha medusula*	17
Chilodonella cucullus	6	*C. octospinus*	61	*Epalxis antiquorum*	5
Dileptus anser	18	*Frontonia leucas*	34	*E. sp.*	21
Frontonia sp.	18	*Holophyra simplex*	53	*Ludio parvulus*	5
Halteria grandinella	59	*Lacrymaria olor*	20	*Metopus campanula*	8
Hemiophrys procera	21	*L. pupula*	52	*M. es*	12
Homalozoon vermiculare	6	*Lembadion bullinum*	39	*M. intercedens*	6
Loxophyllum meleagris	6	*L. magnum*	8	*M. rostratus*	7
Nassula ornata	6	*Lionotus vesiculosus*	17	*M. spinosus*	39
Trachelius ovum	5	*Loxodes magnus*	37	*M. spiralis*	20
Trachelophyllum		*L. striatus*	18	*M. undulans*	16
sigmoides	5	*Loxophyllum helus*	15	*Myelostoma bipartitum*	32
Vorticella convallaria	5	*Paramecium caudatum*	11	*M. sp.*	39
		Pleuronema crassum	44	*Plagopyla nasuta*	14
		Prorodon teres	50	*Saprodinium putrinum*	8
		Spirostomum filum	22		
		S. minus	61		
		Stentor coeruleus	26		
		S. polymorphus	17		
		Urocentrum turbo	21		
		Uroleptus longicaudatus	12		
		Urosoma ceinkowski	13		

CHAPTER 10

The Bottom Fauna: Studies of Communities

In 1963 the author and Miss Rachel Maudsley (now Mrs Hans Erwig) started a survey of the animals inhabiting the stony substratum of Windermere.

Preliminary collections were made in 1963 and 1964 in Windermere and three other lakes, the main unknown at that time being the number of collections necessary to give a reliable picture of the fauna of a lake. The other consideration was time of year but we were less in the dark here as the life history of most, though not all, of the species was known. Most insects grow during the winter and some are present only in the egg stage during the summer, for which reason July, August, September and possibly June are unsuitable for the main collections. There are, however, a few species which are to be taken only then. May and October, when it is possible to work all day, are the best months from the point of view of collectors' comfort; the months in between are the best from the point of view of the presence of animals, but lower temperature limits the time that can be spent collecting. We avoided periods of high level because, though some animals move up when the water rises (Moon 1935b), some do not, and an error may be introduced if a population of constant size should prove to distribute itself over an area that expands and contracts.

During the preliminary years, collecting continued for five minutes at rather few stations, though at enough to show that the distribution of the fauna was not uniform. It became evident that, for the main survey, collections must be made at as many places as possible and that time spent at each one should be short. Accordingly we adopted Moon's technique of collecting for two minutes and then moving on 50 m. Five two-minute collections were regarded as a station. Coarse nets with 10 meshes/cm were used, and we were limited to the depth we could reach with a bare arm. In the North Basin only a few stations were located on the east side because a number of large houses are situated near the edge of the lake. Some householders have altered the lake edge where their gardens come down to it, and there was always the risk that fauna might be affected by the effluent from an undetected septic tank.

May and October 1965 were dry months, and both were devoted to collections in Windermere, where, owing to proximity to our base, we could make a large number of collections each day. During the next two years collections were made in the other lakes, many of them in winter.

140

Observations in Windermere, as nearly once a month as rises in lake level allowed, were continued for a year from October 1965.

Three preliminary points need consideration before the main results are presented and discussed. The first is the variation in numbers from one collection to another, to illustrate which the winter collections from two (3 and 8 see fig. 53) of the five stations visited regularly are set out

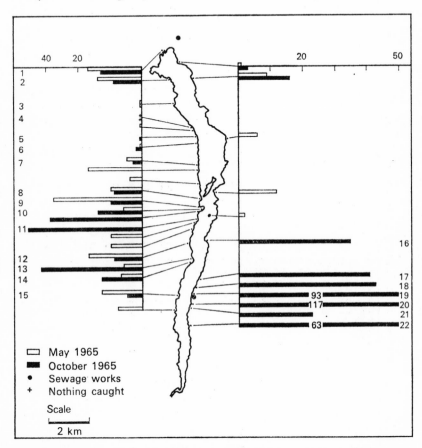

53. The numbers of *Polycelis nigra* and *tenuis* caught at different stations in Windermere.

in table 35. When tested statistically *Ecdyonurus dispar* was found to be randomly distributed in each of the six months, which suggests that the collecting technique was satisfactory. It is not, however, possible to explain why the total in November was so much lower than all the others. The remaining species showed some tendency to clump together, but the range is never so great as to cast doubt on the validity of the total figure for at least rough comparative purposes.

The second point is variation in numbers due to events in the life history.

141

Table 35

Numbers caught in 2-minute collections, at two stations, Epley Point (3) and Coatlap Point (8). Species of which fewer than 5 specimens were taken: *Chloroperla torrentium, Polycelis felina, Glossiphonia complanata, Helobdella stagnalis, Oulimnius tuberculatus, Hydraena gracilis, Planorbis albus, P. contortus* (*Leptophlebia marginata* has been left in the table)

WINDERMERE 2 min colls 50 metres apart

Species	St	OCTOBER 1	2	3	4	5	Total	NOVEMBER 1	2	3	4	5	Total	JANUARY 1	2	3	4	5	Total	FEBRUARY 1	2	3	4	5	Total	MARCH 1	2	3	4	5	Total	MAY (beginning) 1	2	3	4	5	Total
Ecdyonurus dispar	3	9	17	20	8	7	61	6	11	3	4	5	29	17	10	13	12	14	66	5	15	14	15	19	68	2	8	10	7	16	43	12	16	7	10	9	54
	8	3	2	9	13	—	27	—	—	3	1	—	6	11	2	1	3	—	4	—	1	1	3	7	29	—	1	3	—	2	9	2	5	1	2	5	17
Heptagenia lateralis	3	3	2	5	1	3	17	3	2	3	3	4	12	1	—	4	1	—	2	1	1	—	3	1	6	1	1	3	3	7	15	3	1	4	3	4	15
Centroptilum luteolum	8	8	—	5	1	3	—	—	—	—	—	—	4	—	4	—	—	2	—	1	—	—	—	—	—	—	1	—	—	1	5	1	1	—	1	—	3
Leptophlebia marginata	3	2	—	2	—	1	5	—	1	1	—	—	2	—	—	—	—	—	—	—	—	—	—	—	1	—	—	—	—	1	1	—	1	—	—	—	1
Diura bicaudata	3	2	—	1	1	1	5	2	2	2	2	9	23	1	1	1	1	2	7	—	—	1	1	1	2	—	1	—	1	1	—	—	1	—	1	—	—
Nemoura avicularis	8	—	—	9	6	6	44	—	—	3	3	—	4	—	—	1	—	1	2	—	—	—	—	1	—	—	1	1	—	—	1	—	—	—	—	—	—
Polycelis nigra (agg)	3	11	9	12	6	6	44	3	3	7	3	—	16	46	5	20	12	69	152	53	24	49	9	20	155	20	12	12	24	38	86	2	2	27	14	167
Dugesia lugubris	8	—	—	—	—	1	1	—	—	1	1	—	1	3	—	1	1	1	4	1	3	1	1	1	13	1	—	1	—	2	—	—	—	—	—	3
Dendrocoelum lacteum	3	1	—	2	—	—	2	1	—	—	—	1	1	—	—	—	—	—	—	—	—	—	—	1	—	—	—	—	—	—	—	—	—	—	—	—
Erpobdella octoculata	8	1	—	—	—	2	3	—	—	—	—	—	—	3	—	1	1	5	10	1	3	6	2	10	22	—	—	1	—	4	—	1	3	1	—	9
Piscicola geometra	8	1	2	2	2	5	—	—	1	—	2	1	1	—	—	—	1	1	2	—	1	—	—	1	1	—	1	—	2	2	—	—	—	1	1	4
Gammarus pulex	3	2	4	—	7	9	23	1	1	6	2	10	—	1	5	4	1	14	6	2	1	—	1	9	—	4	3	2	8	5	1	1	1	1	8
Crangonyx pseudogracilis	8	5	1	—	1	1	—	—	—	—	—	6	3	4	2	1	—	10	—	1	2	1	12	—	3	3	1	12	—	2	2	1	8
Asellus aquaticus	3	6	8	—	7	2	28	—	3	1	4	13	10	1	16	8	22	57	4	11	22	1	10	48	6	7	3	13	32	1	6	1	6	12	35
Deronectes depressus	5	10	3	4	2	24	13	27	9	20	18	87	4	3	8	6	8	29	3	13	3	1	3	23	8	15	2	8	5	38	4	3	—	24	7	38
Haliplus fulvus	8	—	—	—	2	2	—	—	—	—	—	—	3	—	3	—	—	3	1	—	—	—	1	1	5	—	—	1	1	8
Ancylus fluviatilis	3	2	1	—	4	1	6	19	3	7	3	38	14	28	10	30	2	84	14	9	28	24	2	77	8	18	19	40	16	101	44	19	32	29	10	134
Physa fontinalis	8	1	1	2	3	1	5	2	1	1	5	9	17	9	6	9	1	15	11	1	3	1	16	4	5	6	8	19	42	6	13	21	6	40
Limnaea pereger	8	3	2	—	—	—	3	—	—	1	—	—	1	2	1	—	—	—	—	9	1	—	2	2	12	2	2	2	—	2	12	1	2	1	—	—	2
Agapetus fuscipes	3	91	142	314	314	16	877	76	128	64	57	95	420	243	109	64	78	19	513	142	33	43	70	18	306	92	78	59	37	68	334	65	94	146	92	53	450
	8	223	264	19	20	—	526	315	81	4	51	5	456	124	—	4	5	1	134	1	100	2	25	2	128	33	437	87	14	121	692	45	—	32	4	273	354
Polycentropus flavomaculatus	3	4	4	8	2	9	27	5	6	9	8	1	26	2	4	3	5	7	21	2	4	4	6	2	22	3	4	4	1	11	2	4	9	1	16
Cyrnus trimaculatus	8	1	—	5	1	1	1	—	1	—	—	1	1	—	—	1	—	—	1	—	2	—	3	1	1	1	2	—	1	1	2	1	4

3—Epley Point
8—Coatlap Point

This mainly concerns the insects and is discussed more fully by Macan and Maudsley (1968). In the present context it is sufficient to say that, though some of the Plecoptera started to emerge in May or earlier, *Capnia bifrons* was the only species whose numbers were significantly lower in collections made in this month compared with those immediately preceding. An unusually large haul of *Ecdyonurus dispar* in July was obviously caused by the appearance of a new generation but, in general, marked rises attributable to this were unexpectedly infrequent. Mr George Thompson has informed me in a personal communication that the Freshwater Biological Association's supply department sometimes gather great numbers of an ecdyonurid nymph at the lake's edge at a time when congregation there before emergence is likely. This phenomenon was not observed by us in Windermere, possibly because it is not an annual occurrence, possibly because it is of short duration and we never happened to be collecting at the time. However, one unusually large collection in Derwentwater of *Heptagenia lateralis* may have been caused by this.

The third point is distribution in depth, an important one because the technique used of lifting stones by hand limited operations to water shallower than about 60 cm. In March 1967 the University of Newcastle-upon-Tyne Sub-aqua Club collected at four of the standard stations. At three the stony substratum extended down a little beyond a depth of 240 cm but at the fourth, which was more protected, there was deposition of silt and fine organic matter between 180 and 240 cm. Their results are set out in table 36. *Ecdyonurus dispar* is much more numerous at the water's edge than further out and the same is true of most of the Ephemeroptera and Plecoptera though the preponderance in shallow water is less. Exceptions are *Centroptilum luteolum* and *Leptophlebia marginata*. The latter occurred in the standard collections (e.g. table 35) almost exclusively in May and is the only species that was found by the divers but was consistently absent from most of the standard collections. The species are arranged in descending order roughly according to the extent to which they are more abundant in deeper water, and it is evident that the abundance of some is likely to be greater than is revealed by collections made by a collector not equipped to submerge. It is hoped to repeat this survey to discover whether any of the species move into shallower water in summer, as they do in Esrom Lake (Berg 1938), though there is nothing in the data gathered (e.g. table 35) to suggest a similar movement in Windermere.

In 1964 two collections were made at each 5-minute station, one at a depth of a few centimetres, the other in water where the collector could only just reach the bottom with his arm. The results, as far as they go, agree closely with those obtained by the diving team.

The result of collecting of the kind described is a list of species so long that it is difficult to draw from it conclusions comprehensible to anybody except a worker who is extremely familiar with the particular community

143

Table 36

Numbers at five depths on stony substratum. March 1967. (Fifteen 2-minute collections at 15 and 240 cm and twenty 2-minute collections at 60,120 and 180 cm)

Depth	15	60	120	180	240	cm
Ecdyonurus dispar (Ephem.)	77	5	1	–	–	
Heptagenia lateralis (Ephem.)	9	2	1	1	–	
Nemoura avicularis (Plecopt.)	15	3	4	2	–	
Chloroperla torrentium (Plecopt.)	11	–	–	1	–	
Capnia bifrons (Plecopt.)	11	1	–	–	–	
Ancylus fluviatilis (Moll.)	105	51	41	23	7	
Crangonyx pseudogracilis (Crust.)	87	7	6	11	8	
Polycentropus flavomaculatus (Trichopt.)	3	2	3	–	–	
Physa fontinalis (Moll.)	19	20	22	11	25	
Polycelis nigra (Triclad.)	25	3	12	15	24	
Agapetus fuscipes (Trichopt.)	145	665	300	178	121	
Centroptilum luteolum (Ephem.)	10	14	49	24	22	
Leptophlebia marginata (Ephem.)	–	5	11	5	2	
Valvata piscinalis (Moll.)	–	9	2	9	1	
Gammarus pulex (Crust.)	34	13	22	31	50	
Asellus aquaticus (Crust.)	16	14	51	71	76	
Erpobdella octoculata (Hirud.)	1	9	11	26	26	
Piscicola geometra (Hirud.)	4	4	–	9	1	
Glossiphonia complanata (Hirud.)	1	1	5	11	9	
Dugesia lugubris (Triclad.)	–	–	4	10	7	
Dendrocoelum lacteum (Triclad.)			2	17	2	

studied. It is desirable, therefore, to shorten the list as much as possible before discussing it. A common practice is to discard all species which do not attain a certain abundance, but this is open to the objection that it eliminates species which, though rare, are characteristic of the community. An extreme example is the helminthid beetle, *Stenelmis canaliculata*, which was recorded in Windermere by Claridge and Staddon (1961) and has not been found anywhere else in the British Isles. Macan and Maudsley (1968) record forty-eight species of insect in stony parts of Windermere, but retain only fourteen of them as typical inhabitants of the stony substratum. The rest are discarded on the grounds that their numbers are so much higher in neighbouring biotopes, in sand, in mud, in vegetation or in the inflowing streams, as to suggest that their presence on the stony substratum is caused by migration from those biotopes. The arbitrary rejection is perhaps excessively drastic, but, until more reliable criteria are available, probably justified by the resulting clarity of the general picture.

In the other groups, *Polycelis felina* has been omitted from the Platyhelminthes because, although it is numerous in places, there is reason to believe that it would not survive permanently but for reinforcements from streams (Beauchamp 1932).

As noted in the preceding chapter only the three species of mollusc in the list below belong to the stony-substratum community.

There seems no good reason to exclude any leech.

The list, therefore, is:

Plecoptera
Nemoura avicularis (Morton)
Diura bicaudata (Linn.)
Chloroperla torrentium (Pict.)
Leuctra fusca (Linn.)
Capnia bifrons (Newm.)

Ephemeroptera
Ecdyonurus dispar (Curt.)
Heptagenia lateralis (Curt.)
Centroptilum luteolum (Curt.)

Trichoptera
Polycentropus flavomaculatus (Pict.)
Agapetus fuscipes (Curt.)
Cyrnus trimaculatus (Curt.)
Tinodes waeneri (Linn.)

Coleoptera
Platambus maculatus (Linn.)
Stenelmis canaliculata (Gyll)

Crustacea
Gammarus pulex (Linn.)
Crangonyx pseudogracilis (Bousfield)
Asellus aquaticus (Linn.)
Asellus meridianus (Rac.)

Mollusca
Limnaea pereger (Müll.)
Physa fontinalis (Linn.)
Ancylus fluviatilis (Linn.)

Platyhelminthes
Polycelis nigra (Müll.)
P. tenuis (Ijima)
Dugesia lugubris (Schmidt)
Dendrocoelum lacteum (Müll.)
Bdellocephala punctata (Pall.)
Planaria torva (Müll.)

Hirudinea
Erpobdella octoculata (Linn.)
Glossiphonia complanata (Linn.)
Helobdella stagnalis (Linn.)
Glossiphonia heteroclita (Linn.)
Piscicola geometra (Linn.)
Theromyzon tessulatum (Müll.)
Hemiclepsis marginata (Müll.)
Haemopis sanguisuga (Linn.)
Batracobdella paludosa (Carena)

Hydrachnellae, oligochaetes, chironomids and other Diptera, some Trichoptera, and animals too small to be retained by the coarse net used, have been ignored.

There are thirty-six species in this list but some have been taken only rarely and in small numbers. There are nineteen of which this is not true and on which the discussion of the communities is mainly based. Their distribution is shown in figs. 53 and 54. Fig. 53 is a map of Windermere on which the position of all the collecting stations has been marked. Those to which reference is made in the text are numbered serially along the margin of the diagram. The histograms show the numbers of *Polycelis* taken at each station, *P. nigra* and *P. tenuis* being treated as one on account of the excessive consumption of time involved in distinguishing them. Both March and October collections are shown. The numbers in the two months vary considerably and at no station were very large numbers taken twice. What does come out of the figure is that numbers were always low in the middle part of the North Basin from Epley Point (st. 3) to Strawberry Gardens (st. 7) just to the north of the island region, and almost always high everywhere else. The October figures, repeated on fig. 54, show a gradual fall and then a gradual rise with progress from north to south.

Above what is here referred to as *Polycelis nigra* agg., a convenient if not strictly accurate adaptation of botanical nomenclature, there are in fig. 54 five species with a similar distribution: two more planarians, an isopod, an amphipod and a snail. They form the nucleus of what may be called, since none of them is an insect, the non-insect community, although some representatives of that class do form part of it. Next in order up the left-hand column comes *Gammarus pulex*, which, though more abundant in the south than in the north, as are the preceding, is more uniformly distributed in the North Basin than any of them are. Unlike them, it was, in October, scarce at or absent from stations 7, 8 and 9, the first of which is near, and the next two in the particularly sheltered area of the islands. *Limnaea pereger* is widely distributed but more abundant in t he South Basin than in the North. *Ancylus fluviatilis* is probably uniformly distributed on the stony substratum in the lakes, but patchily, so that a large total depends on the collector having by chance encountered a patch where numbers are high.

The stonefly, *Diura bicaudata*, at the base of the right-hand column of fig. 54 occurs only where the members of the non-insect group are scarce and was never recorded except at stations 3, 4, 5 and 6. *Ecdyonurus dispar* was a little more widespread and was recorded as well at stations 2 and 7 in May. The remaining Ephemeroptera and Plecoptera clearly form a group with the two already mentioned, though they are found at a few stations in other parts of the lake. The distribution of *Polycentropus flavomaculatus* was similar except that high numbers were often found in the South Basin. The October collection was unusual in that specimens were few at station 10 (Macan and Maudsley 1968). This station lies on the south side of the promontory on which the Ferry House is situated, and, lying at right angles to the long axis of the lake, is one of the most exposed. The next comparable promontory is Rawlinson's Nab, and *Polycentropus* occurred on the exposed south side of this at station 14 but not on the sheltered north side at station 13. The next projection, Grubbins Point, is smaller but it is a headland and the largest number of *Polycen-*

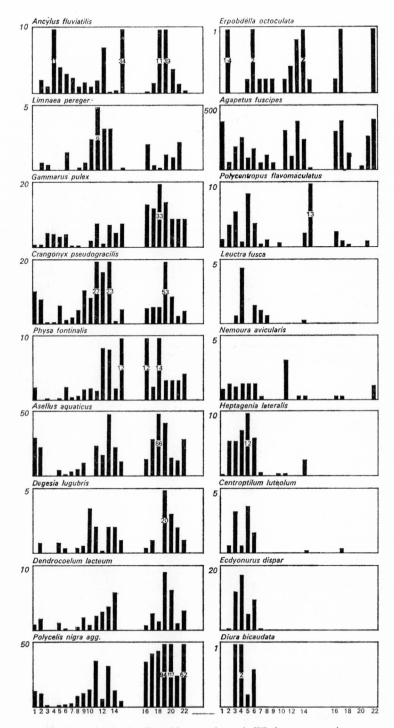

54. Numbers of animals collected in two minutes in Windermere at stations numbered as in fig. 53. The numbers are the average of ten, sometimes five, collections.

tropus in October was taken here. Station 11 is a small point between Ferry House and Rawlinson's Nab. It was at these exposed stations that the other members of the insect group were generally taken outside the North Basin. *Agapetus fuscipes*, by far the commonest insect, shows some affinities with the non-insect group, but is scarcest at stations 7, 8, 9 and 10 and 19 and 20. The allocation to each species of the same amount of space in fig. 54 with corresponding adjustment to the scale gives a false impression of the distribution of *Erpobdella octoculata*, whose apparently large peaks represent two specimens only. Consideration of all the records indicates that it was uniformly distributed. Two other leeches, *Helobdella stagnalis* and *Piscicola geometra* appeared to belong to the non-insect community, but the numbers taken were too small to justify a definite statement.

There are then two distinct communities, and five species that fall less clearly than the rest into either. The non-insect group occupies the extreme north end of the lake but is scarce along the west shore in the middle of the North Basin. It becomes abundant again as the islands are approached and persists in the South Basin. The insect group is abundant where the other is scarce and largely confined elsewhere to the most exposed places. *Ancylus fluviatilis* and *Erpobdella octoculata* are uniformly distributed. *Gammarus pulex*, *Limnaea pereger* and *Agapetus fuscipes* also have a more uniform distribution than most species, though they show affinities with the non-insect group.

The work has not passed the descriptive stage, but it is legitimate to look for factors with which the occurrence of the two groups could be correlated. There is one obvious one—enrichment by sewage. The Ambleside sewage works discharge into the main inflow into the North Basin, and it is reasonable to assume that the prevailing southerly wind carries the dilute effluent on to the north shores of the lake. Further south the west shore of the lake is the least inhabited part and the volume of water is greater than in the South Basin. In the region of the islands there are houseboats in summer, and Ferry House has its own septic tank which discharges straight into the lake. Possibly other establishments have similar septic tanks. The South Basin, it is believed, is affected by the outfall from the Bowness–Windermere sewage works. It is noteworthy, in support of this contention, that the two largest collections of *Polycelis nigra* were taken on either side of the outfall of these works. Five hundred and eighty-seven were taken on the south side and 464 on the north but here there was a small stream entering the lake which no doubt accounted for the 861 *Polycelis felina* taken as well. Reynoldson (1966) has shown that these two species have similar requirements, and it is therefore justifiable to add their numbers together, which gives a total of 1326. Moreover, Macan (1962b) observed a great increase in the number of *Polycelis felina* in a small stony stream when the load on a septic tank that discharged into it was increased.

The biggest catches of all the other main members of the non-insect group and of *Gammarus pulex* were made near the outfall of the Bowness–Windermere sewage works, generally on the north side. The same was true of *Ancylus fluviatilis*, but numbers of *Agapetus fuscipes* were conspicuously low in the immediate vicinity of the outfall.

The association between *Asellus* and pollution is well known, but generally the other abundant animals are leeches not flatworms (Hynes 1960). This, however, is usually in the zone of recovery from gross pollution, conditions which are not found in Windermere. Shortly before this work was started, the Ambleside sewage works were rebuilt and an activated sludge plant installed. The effluent from the Bowness–Windermere works was less satisfactory, and during our collecting operations, the works were being enlarged by the building of additional sprinkling filters. Even an unsatisfactory effluent is greatly diluted in the lake, and it is clearly advisable to speak of enrichment rather than pollution. How lake water enriched in this way provides the conditions in which flatworms, crustaceans and snails flourish is unknown.

The striking curtailment of the range of most insects remains to be examined. No appreciable lowering of the oxygen concentration has been observed, and it is unlikely that the effluents bring in substances harmful to insects but not to flatworms. Macan (1965b) has tentatively put forward the following explanation. Though Ephemeroptera and Plecoptera can obviously not be the main food of flatworms, they are eaten by them. When enrichment by sewage increases the chief food supply (whatever it may be) of flatworms sufficiently for them to become numerous, their predation on insects reduces the number of the latter. Further it is likely that omnivores, such as *Gammarus* and *Asellus*, eat large numbers of mayfly and stonefly eggs, and these may even be consumed by herbivores such as snails as they glide across the stones rasping a swathe through the algal felt. A supporting observation is that the eggs of Ephemeroptera and Plecoptera generally fall to the bottom haphazardly, whereas the animals that are successful in crowded conditions produce eggs that are protected. Those of flatworms and leeches are enclosed in a tough cocoon; those of snails in a mass of jelly; and *Gammarus* and *Asellus* carry theirs until they hatch. The suggestion is then that under productive conditions, where these carnivorous and detritus-feeding animals abound, certain insects are absent because they are not adapted to meet the resulting predation.

Why the insects mentioned can survive this predation on the most exposed shores is not known. It is pertinent to remark in this connexion that this is the sort of stumbling block which is too often ignored by the advocates of the 'Saprobiensystem', which is still popular on the Continent. This seeks to define the range of each species in terms of pollution alone. The author's belief is that the occurrence of any species depends on so many factors that no attempt to explain it in terms of one can be satisfactory.

Table 37 shows the occurrence of most of these species in the other lakes, the figures being the number collected in 100 minutes, and the order being designed so that species most abundant in the productive lakes come first and the peak of abundance of the rest shifts steadily to the right. Neither of the two beetles were taken in these collections, and not all the Platyhelminthes or Hirudinea are included as both groups are discussed separately later. Most of the collections were made during the winter and few in July, August and September. Fortunately regular collections in three lakes were made throughout the summers of the early years and study of these showed that *Ecdyonurus dispar* appeared in May, abounded in June, July and August and was not to be found thereafter. Records for September, however, are few and the season may extend into that month. Surprisingly large populations were found in lakes where extensive collecting in winter failed to reveal the species. This was unexpected because in Windermere *E. dispar* occurs in every month with slight variation in numbers (Macan and Maudsley 1968). The same is probably true in Ullswater. In Windermere there is a long over-wintering generation and a short summer generation; the life history in the other lakes is under investigation. Another unexpected finding is that *E. dispar* is also a summer species in Esthwaite. The figures in table 37 give therefore no true picture of its abundance, but its variable life history was noticed in time to organize summer collections in 1967. These were not sufficient to justify quoting numbers but did establish that it is widespread, the only lake in which it was not found being Coniston. The one wholly summer species is *Leuctra fusca*, which has not been recorded in Derwentwater, Bassenthwaite, Ullswater and Coniston.

The total number of organisms taken in each lake is shown at the bottom of table 37. *Agapetus* has been excluded on account of its erratic occurrence and great numbers when it does occur. The highest total, by a big margin, is in Esthwaite and it may be regarded with confidence on account of the uniform distribution of species in the lake. The last five lakes show a serial decline that accords well with what is already known about them, but in between come six lakes whose totals do not coincide with the order in which the lakes have been arranged.

The figures are repeated in the form of histograms in fig. 55. *Polycelis nigra*, as might be expected from the Windermere records, ranges from very numerous in the most productive lakes to sparse or absent in the least. *Asellus* is another organism plentiful in productive lakes but it has been omitted from the figure because absence may be due to failure to reach a lake rather than to unfavourable conditions. *Gammarus pulex* is similar to *Polycelis*, but its fall in the first five lakes is much slower and it is scarce only in the last three. (Thirlmere may be ignored for the moment.)

Numbers collected on stony substratum in 100 minutes

	Esthwaite	Windermere S.	Bassenthwaite	Coniston	Loweswater	Windermere N.	Ullswater	Derwentwater	Crummock	Buttermere	Ennerdale	Wastwater	Thirlmere
Polycelis nigra	2830	907	356	32	222	11	6	1		3			
Asellus aquaticus	1782	579	222			18							
Asellus meridianus		11				1							
Gammarus pulex	775	274	618	537	710	95	395	147	159	+	3		14
Crangonyx pseudogracilis		228	150			33	25						
Dugesia lugubris	27	35	58	1		10	+	7					
Dendrocoelum lacteum		40	1			4							
Helobdella stagnalis	7	11	20	1	17	2		12	1	2			
Physa fontinalis	19	149	339	128	2	8	3		18				
Polycentropus flavomaculatus	109	86	139	11	179	143	56	55	16	107	24	40	
Erpobdella octoculata	61	26	19	32	60	9			4	23	20	26	
Glossiphonia complanata	2	5	9	10	7	3	1	4	2	3	4		
Agapetus fuscipes	53	4285	1	2	1099	4128	1894	203	2				
Ancylus fluviatilis	322	361	170	388	1302	899	298	15	128	112	177	5	
Cyrnus trimaculatus	5	6	60	7	13	11	15		3	3	1	32	
Nemoura avicularis	21	3	18	24	3	38				5	1		9
Limnaea pereger	3	24	262	61	65	24	95	267	40	5	76		
Centroptilum luteolum	1	3	9	22	11	78	16	24	5		13		23
Heptagenia lateralis		46		4		210	25	402	226	23	44	13	
Ecdyonurus dispar		8			111	480	689	1	14	22		4	
Chloroperla torrentium	23	13	4	13	24	22	23	110	90	87	53	26	
Diura bicaudata			54	16	13	25	49	54	80	50	86	40	
Capnia bifrons				7		15			61		7		
Leuctra fusca												13	2
Total (without *Agapetus*)	5987	2815	2508	1294	2739	2129	1696	1307	847	455	515	194	

The next three species in fig. 55, that is a flatworm, a leech, and the snail *Physa*, are all numerous only in productive lakes, but are distinctly scarcer in Esthwaite than in some others. The remaining species all have a more uniform distribution in all the lakes. The most difficult occurrence to explain is that of *Agapetus*, of which thousands were taken in Windermere, Loweswater and Ullswater, fifty-three in Esthwaite, a few in four other lakes including Wastwater, and none in the rest. *Diura bicaudata*, the last species, was absent from Esthwaite as well as from the sewage-influenced parts of Windermere, and most abundant in the lakes at the unproductive end of the series though the numbers in Wastwater were small. *Chloroperla torrentium*, *Heptagenia lateralis* and *Centroptilum luteolum* are all similar, though the point of greatest abundance is progressively shifted in the direction of the productive lakes. On the other hand what was shown in Windermere to be the fifth typical member of the 'insect group', *Nemoura avicularis*, is uniformly distributed in most of the lakes on the productive side of the halfway line and scarce or absent on the other.

Another approach to these data is to take the lakes one by one and compare them with others. Esthwaite and that part of Windermere which is enriched by sewage clearly come together in a group characterized by abundance of flatworms and *Asellus* and a paucity of Ephemeroptera and Plecoptera. Certain differences, notably the greater number of the first two, and the absence of *Heptagenia lateralis*, are likely to be related to the more productive conditions in Esthwaite. To offer an explanation of the smaller numbers of *Agapetus* and *Physa* would be rash until more is known of the interactions of the various species. The absence from the standard collections of *Dendrocoelum*, which has, however, been recorded on other occasions, is unexpected.

At the other end of the series Buttermere, Ennerdale and Wastwater fall into a group characterized by relative abundance of Ephemeroptera and Plecoptera and small numbers of species that are not insects. The year of collection was an outstandingly successful one for *Limnaea pereger* in Ennerdale, but apart from this, there is a tendency for numbers in all groups to be lowest in Wastwater, which in this field, appears to be less productive than Ennerdale and Buttermere. No scheme shows much difference between the three but Ennerdale and Wastwater are generally regarded as equal and least productive, with Buttermere a trifle above them.

Derwentwater and Crummock form another group distinguished from the foregoing by the occurrence of *Physa fontinalis* and the greater numbers of *Heptagenia* and *Gammarus*.

This leaves four lakes, Bassenthwaite, Coniston, Loweswater and Ullswater, that fit less satisfactorily into any serial arrangement. Bassenthwaite is the most like Windermere S. and Esthwaite but differs in that the

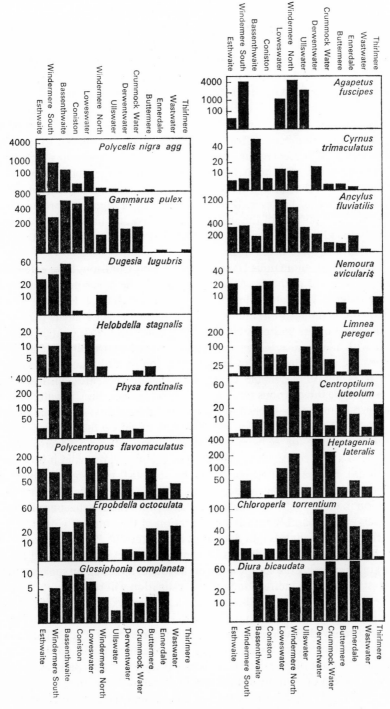

55. The numbers of animals collected in various lakes in 100 minutes.

153

population of *Agapetus* is very small. Coniston has been placed next to it to bring all the lakes with few *Heptagenia* together. It also has scarcity of *Agapetus* in common with Bassenthwaite. On the other hand on the basis of the number of flatworms it should be further to the right. Loweswater with a fair population of both flatworms and *Heptagenia* falls obviously into a middle position. With the lakes arranged in the order used in table 37 it is the first, reading from the left, in which *Physa* is scarce. On the other hand the large numbers of *Agapetus* indicate a position further to the left. Ullswater has few *Heptagenia* but great numbers of *Ecdyonurus* which may be a competitor. If it be assumed that it is, the large numbers may be correlated with the paucity of flatworms and the lake is correctly placed in fig. 55. Here too, however, the big population of *Agapetus* provides an anomaly. The poverty of the snail fauna of Ullswater has been mentioned already (p. 129) and a glance at table 37 shows that fewer leeches than in any other lake were taken in it.

Two of these anomalous lakes, Ullswater and Coniston, have extensive mine spoil-heaps in the drainage area. Though the concentration of lead in the water of Ullswater is low, in the mud it is high (F. J. H. Mackereth: personal communication) and it could from here affect animals that feed on attached plants and possibly those that come into close contact of any kind with the surface. Carpenter (1926) showed that molluscs and flat-worms are eradicated at a lower concentration of lead than most insects, and it is not unreasonable to suppose that leeches are sensitive too. These are the three groups that are unexpectedly scarce in Ullswater and it does not seem too far-fetched to suggest that their scarcity and lead might be related. The low predation by flatworms might be the reason for the great abundance of ecdyonurids which either suffer no adverse effect from lead at the concentration which they encounter, or avoid it by living on stones, not on the bottom. If this be accepted, the large numbers of ecdyonurids might indicate that Ullswater is more productive than its position in table 55 shows, which is in accord with the conclusions of other workers. A comparably plausible explanation of the fauna of Coniston, in which both ecdyonurids and flatworms are scarce but *Physa* numerous, is much more difficult. If pollution by some heavy metal from the old mine workings is altering the composition of the fauna, its action on some species at least is probably indirect and only investigation will reveal it.

Yet a third approach to these data is a detailed examination of certain groups, paying to the less common species more attention than they have hitherto received. Table 38, showing the occurrence of Platyhelminthes, is based on the records already quoted, on collections made at other times and on the work of Karen Clarke, Susan Hargreaves and Barbara Welch who, in 1962, visited most of the lakes in search of flatworms. It was they who undertook the laborious task of squashing every specimen so that *Polycelis tenuis* could be distinguished from *P. nigra*. Ennerdale is shown

Table 38

Occurrence of *Platyhelminthes* in the Lake District lakes

	Polycelis nigra	P. tenuis	Dugesia lugubris	Dendrocoelum lacteum	Bdellocephala punctata	Planaria torva	Fonticola albissima
Wastwater	X						
Ennerdale							
Buttermere	X	X				X	
Crummock	X	X					
Derwentwater	X	X					
Bassenthwaite	X	X	X	X			
Coniston	X	X	X	X	X		
Loweswater		X					
Ullswater	X	X		X			
Windermere	X	X	X	X	X		
Esthwaite	X	X	X	X	X		
(Esrom)	X	X	X	X	X	X	X

as without flatworms; in fact we once took one small one there but did not determine to which of these two species it belonged.

Only one species in Loweswater is unexpected. The poor fauna of Ullswater has already been discussed. Otherwise the number of species declines with the decrease in productiveness just as that of the molluscs (table 29) did.

The records of Hirudinea treated in the same way give a similar result (table 39).

Any discussion of lake classification in terms of these findings must begin with an examination of the validity of the whole concept. If, along a stretch of lake shore, there is a change from one community to another with no difference in the physical conditions, it is questionable whether much is to be gained from treating the lake or lake basin as a unit. One may indeed go further and suggest that, if differences in the nutrient supply can affect the littoral communities so drastically, they must also affect the plankton communities. Here, however, differences are continually being blurred or erased by the wind. The idea is not new. Valovirta (1958) recorded a steady diminution in the amount of bottom fauna down the length of a long Finnish lake of which the main inflow was a river entering

Table 39
Occurrences of *Hirudinea* in Lake District lakes

	Erpobdella octoculata	*Glossiphonia complanata*	*Helobdella stagnalis*	*Glossiphonia heteroclita*	*Piscicola geometra*	*Theromyzon tessulatum*	*Hemiclepsis marginata*	*Erpobdella testacea*	*Haemopis sanguisuga*	*Batracobdella paludosa*
Wastwater	X									
Ennerdale	X	X	X							
Buttermere	X	X	X	X						
Crummock	X	X	X			X				
Derwentwater	X	X	X							
Bassenthwaite	X	X	X	X						
Coniston	X	X	X						X	
Loweswater	X	X	X			X				
Ullswater		X	X	X						X
Windermere	X	X	X	X	X	X			X	X
Esthwaite	X	X	X	X	X	X	X		X	
(Esrom)	X	X	X	X	X	X	X	X		

at one end. This effect will be apparent only in a lake above a certain size, and will be more marked if one inflow is strongly enriched from sewage or some other source.

Esthwaite, the smallest of the lakes in table 37, had an unusually uniform fauna in spite of considerable enrichment at one end only. Presumably in a basin as small as this circulation is sufficient to mix the inflow water with the rest of the lake too quickly for organisms at the inflow end to take advantage of their proximity to the source of supply. The other lakes are either small or not greatly enriched by sewage and in the Lake District, at least, the combination of considerable local enrichment and large size is not sufficiently frequent to make the comparison of whole lakes unprofitable. On total number of organisms collected in a given time there is a clear series from Esthwaite to Windermere and then, after a gap, from Derwentwater through Crummock, Buttermere and Ennerdale to Wastwater, which agrees well with arrangements based on other factors. A study of all the animals, based on fig. 55, leads to the same conclusions. In both schemes the gap between Windermere and Derwentwater is occupied by four lakes which, though not outstandingly anomalous, do not

156

fit into the series as well as the others. However, if the suggestion that the fauna of two of them is influenced by pollution from old mine spoil-heaps is shown to be correct, the irregularity in the series will be less. A serial relationship between the lakes appears most clearly when the number of species of mollusc, flatworm and leech in each lake is compared. Although the bottom fauna reflects local conditions much more than plankton or nekton it turns out to be a surprisingly good indicator of the general conditions in a lake.

Information of the kind presented here is an essential prelude to the study of productivity. For studies of the structure of the communities and an explanation of the differences between them, it may prove to be less important. If two communities can be found in one lake, that is the obvious place in which to study them, but there are likely to be occasions when it is advantageous to be able to study one or the other isolated in a lake where only it occurs.

Having put forward the idea that a large superstructure of theory about the community in general has been erected on a small foundation of fact (Macan 1963, p. 16), it is with hesitation that the author contemplates adding to the superstructure on the basis of the findings just presented. The method of collecting was crude and designed to discover no more than the broad outlines of the pattern; moreover it passes over the smallest animals. Nonetheless it does support the idea put forward in two places (Macan 1961b, p. 589, 1963, p. 15) that animal communities grade into one another without the abrupt transition that is seen between plant communities. There is nothing in the animal kingdom that corresponds with the dominant plant, which, having established itself, imposes a pattern on all the other species. In fig. 54 the numbers of *Polycelis nigra* agg. rise regularly from station 4 to station 11. Station 2 has a fauna of inter-mediate type with both *Polycelis* and *Heptagenia* present in moderate numbers. It was a station of ten collections and therefore covered a stretch of shore some 500 m long. Examination of the collections shows that both *Polycelis* and *Heptagenia* were distributed with unusual uniformity, and the similar totals were not made up of a few large hauls from stations at which the two were never abundant together. However, conclusive evidence of a gradual change, rather than the '*mosaïque de petites synusies*' which Marlier (1951) postulates, will be available only after a few critical areas, such as station 2 just mentioned, have been subjected to a more thorough examina-tion over a period of years.

A preliminary survey cannot contribute much towards an explanation of the facts observed. It is, however, possible to offer a partial explanation of the present data because flatworms, which are important members of one of the communities, have been studied for a long time by Dr T. B. Reynoldson and his school. The results of many years devoted to various facets of the problem have recently been the subject of a brilliant synthesis

(Reynoldson 1966). If a series of lakes is arranged according to the concentration of calcium, it is seen that the number of both species and
individuals of flatworms rises with the calcium content. The relationship,
however, is not direct. Flatworms imprisoned in a lake with very soft water
(0·5 mg/l Ca) survived well. Young were reared successfully in the same
water in the laboratory. The relationship appears to be an indirect one
through the food. *Polycelis nigra, P. tenuis, Dendrocoelum lacteum* and
Dugesia (Planaria) lugubris, all feed extensively on small oligochaetes, but
Dugesia takes more molluscs and *Dendrocoelum lacteum* more crustaceans,
especially *Asellus*, than any of the others. In a productive lake there is
food enough for a large population and all four species co-exist because
the feeding habits are not identical. The relationship between the two
species of *Polycelis* remains obscure. In less productive lakes there is less
food and therefore fewer flatworms, which, by growing and reproducing in
times of plenty and reverting to immaturity and diminishing in size when
food is scarce, maintain a steady balance between biomass and food supply.
Not only is there less food, but there is a less varied diet, *Asellus* and
Mollusca being notably absent from or scarce in unproductive lakes.
Whether they, in their turn, are absent because of poor food supply rather
than because of difficulties over ionic uptake, as the flatworms are, remains
to be established. Whatever the cause, the result of the absence or scarcity
is direct competition between all four species of flatworm and in this
scramble for limited resources *Dugesia* and *Dendrocoelum* are not successful. The results from the Lake District lakes (table 37 and fig. 55) illustrate
this relationship well and are very like those quoted by Reynoldson.

It is unlikely that any species that is abundant in an unproductive lake is
prevented from achieving great abundance in a more productive lake by
anything except some kind of interaction with species that are numerous
only in the productive lake. The Ephemeroptera and Plecoptera at the
bottom of figure 55 are scarce on the left-hand side of the figure, that is in
the productive lakes, because they are preyed on, a postulate that was put
forward earlier when the fauna of Windermere was described.

Thirlmere has been included in the table and figure but not so far
mentioned in the text. It might be expected, were it not a reservoir, to hold a
fauna similar to that of the last three lakes, and any differences may be
attributed to the greater rise and fall. Flatworms would not occur anyway
and this is probably true of leeches, for the populations in Ennerdale and
Wastwater appear to be associated with local pollution of which there is
none in Thirlmere. On the other hand the absence of snails, particularly
Ancylus, net-spinning Trichoptera and ecdyonurids suggests that these
animals are unable to keep pace with a falling level. Stoneflies apparently
can, and four species were recorded in addition to the two shown in table
37. Of the Ephemeroptera, *Centroptilum* and also *Leptophlebia* were

comparatively numerous. In Windermere both these species, in contrast to the ecdyonurids, were found to be more abundant in deeper water than in shallow.

There are a few more species found in the various lakes which should, perhaps, be named because they were recorded often, though their numbers were much greater in some neighbouring biotope, generally an inflowing stream. *Polycelis felina*, sometimes very abundant on the lake shore, was the most noteworthy. Many catches contained several species of Plecoptera of which *Leuctra hippopus* and *Amphinemoura sulcicollis* occurred often. *Perlodes microcephala* was relatively numerous in Buttermere and Crummock, apparently living alongside *Diura bicaudata*, which is the only large stonefly in most lakes. A comparable species among the Ephemeroptera is *Ephemerella ignita* which, recorded occasionally in Ennerdale and Windermere, is numerous in parts of Ullswater. This irregular distribution demands study to discover whether it can be related to some peculiarity of the lake or whether it is due to some chance. Should much chance variation in the composition of communities be found to occur, attempts at description and definition, such as has been made above, may prove of little value. That it has been possible to advance a plausible explanation of the distribution of a number of species suggests that chance is not playing a large part.

Micronecta poweri has been mentioned already. Trichoptera have been considered in part because only a few larvae can be identified. Larvae of *Tinodes* have been found in most lakes and the only species taken on the wing is *T. waeneri*. A manuscript key to the larvae of the family to which this species belongs has recently been received from Dr J. Edington but not unfortunately in time for use in the present survey. Beetles in the genera *Haliplus* and *Deronectes* occur on the stony substratum, as do several Helminthidae, of which the most frequently encountered are *Elmis mougetii* and *Oulimnius tuberculatus*.

Reed-beds have received much less attention than the stony substratum. Moon (1936) surveyed one in Windermere and Macan has made extensive collections in connexion with the work on Hemiptera, Mollusca and Coleoptera already mentioned. Of the other groups, Ephemeroptera are represented by both species of *Leptophlebia*, which are often found in large numbers, and by *Siphlonurus lacustris*, which is fairly numerous in the reed-beds at the north end of Esthwaite. It is found also on stony substrata, though sparsely, and has a wide distribution in the lakes. In a lagoon formed by the delta of Wastwater's main inflow it has been recorded in abundance. The ecdyonurids are not found in the reed-beds, nor are most Plecoptera except *Nemoura cinerea*, for which they are the typical habitat. Three species of Odonata, a group almost confined to still water, have been recorded: *Enallagma cyathigerum*, *Ischnura elegans* and *Coenagrion puella*. Moon records *Gammarus* and *Asellus*, *Polycelis nigra*, *Dugesia lugubris*,

Dendrocoelum lacteum, *Glossiphonia* sp. and *Erpobdella octoculata* (*atomaria* in his list). The numerous Trichoptera and Diptera are still unidentifiable.

THE FAUNA OF DEEP WATER

Dr J. H. Mundie worked for some years on the fauna of the deeper parts of Esthwaite, but none of his findings have been published except for a short report to the twelfth Congress of Entomology. In this he states that traps designed to catch both ascending and descending animals have revealed a hitherto unsuspected amount of swimming on the part of chironomid larvae, and also other animals, formerly believed to spend all the time in the mud. This means that species normally found below the thermocline are not cut off from oxygen throughout the period of stratification. Excursions into the water are not, however, confined to species living in the hypolimnion. Swimming by very young larvae seeking an optimum substratum is also mentioned. The main information about the fauna in deep water available is still that of Miss (now Professor) Carmel Humphries who investigated Windermere in the early days. Humphries (1936) started at 3 m, which is near the lower limit of rooted vegetation, and writes that 'collections were made as far as possible near plants but not directly on them, as the Petersen grab did not work satisfactorily over them'. Thereafter samples were taken with the same instrument at intervals of 3m down to 12 m and less regularly with a dredge down to 60 m.

Humphries submitted her specimens of *Pisidium* to Mr A. W. Stelfox and Mr C. Oldham, two well known authorities. Her adult chironomids were identified by Dr F. W. Edwards at the Natural History Museum, and she also received assistance with this group from Professor A. Thienemann and Professor F. Lenz while working subsequently at Plön. She did not, however, succeed in relating all the larvae found to adults, particularly in the Tanypodinae and Tanytarsariae. No reliance can be placed on the names given to the Trichoptera, and there is also reason to believe that the oligochaetes and water-mites were incorrectly named. These last three groups have, therefore, been omitted from table 40, together with those animals which were not identified to species.

It is evident from table 40 that a number of shallow-water species extend down to 3 m. The complete list also contains two dragon-flies, though one is certainly misidentified. Three caddis larvae, all probably of the family Leptoceridae, reached greater depths than this, one going down to 6, another to 9 and the third to 12 m. *Valvata piscinalis* extended down to 9 m. The remaining species at these depths were probably all inhabitants of the mud. A species of *Caenis*, which, from the later work of Kimmins (1954), can with some confidence be identified as *C. horaria*, was found down to a depth of 9 m. *Ephemera danica* had a similar distribution, but *Sialis lutaria* reached 12 m, which would have taken it to a distance from

Table 40

Distribution of animals at different depths in Windermere
The data are taken from Humphries' (1936) table 5, which gives numbers
per square metre in the sublittoral zone and table 2 which indicates presence (X) and absence (–) only
(1) Later work has shown that *L. marginata* is commoner in deep water
than *L. vespertina*. No method of separating nymphs of these two species
was available to Humphries

	3	6	9	12	20	40	60 m
Limnaea pereger	X	–	–	–	–	–	–
Physa fontinalis	2	–	–	–	–	–	–
Planorbis contortus	4	–	–	–	–	–	–
Erpobdella octoculata	12	–	–	–	–	–	–
Glossiphonia complanata	X	–	–	–	–	–	–
G. heteroclita	X	–	–	–	–	–	–
Asellus meridianus	X	–	–	–	–	–	–
Gammarus pulex	16	–	–	–	–	–	–
Leptophlebia vespertina[1]	X	–	–	–	–	–	–
Psectrocladius	X	–	–	–	–	–	–
Chironomus albipennis	X	–	–	–	–	–	–
Protanypus morio	X	–	–	–	–	–	–
Limnochironomus nervosus	18	8	–	–	–	–	–
Valvata piscinalis	25	12	3	–	–	–	–
Caenis	144	5	X	–	–	–	–
Ephemera danica	9	X	X	–	–	–	–
Chironomus cingulatus	22	10	X	–	–	–	–
Pseudochironomus prasinatus	55	18	2	–	–	–	–
Sphaerium corneum	39	15	10	2	–	–	–
Sialis lutaria	27	6	9	1	–	–	–
Pentapedilum tritum	45	24	10	4	X	–	–
Pisidium lilljeborgii	48	48	21	6	X	X	–
P. nitidum	19	4	3	X	X	X	–
P. hibernicum	9	1	1	X	X	X	–
Endochironomus nymphoides	X	X	X	X	X	X	X
Chaoborus plumicornis	–	–	X	X	X	X	–
Stictochironomus rosenscholdi	–	–	X	X	X	X	–
Monodiamesa bathyphila	–	–	–	–	X	X	X
Pisidium casertanum	–	–	–	–	X	X	X
P. personatum	–	–	–	–	X	X	X

the shore which is surprising in view of the facts that it pupates on land and lays its eggs on emergent vegetation. Four species of water-mite were taken in water deeper than 3 m, but beyond 12 m the fauna was made up entirely of chironomids, *Chaoborus*, oligochaetes and *Pisidium*.

Stressing that the figures are approximate, seasonal variation being great, Humphries gives the following as the average number of animals per square metre:

3 m	6 m	9 m	12 m
550	230	100	30

At greater depths the number is thought to be about the same as at 12 m.

Thienemann (1954, p. 437) comments that the chironomid fauna of Windermere recalls that of the oligohumic and oligotrophic lakes of mid and southern Sweden. Numbers per unit area, however, are considerably less (Thienemann 1954, p. 693). The total fauna in the deep water of Windermere is notably sparse when comparison is made with other lakes. The information available provides no explanation of this. It must be stressed that Humphries had but a short time in which to pioneer work that involved much sieving, a particularly time-consuming process. Comment must await a further more extensive programme with samples taken at intervals along the long axis of the lake. Valovirta (1958) has shown that bottom fauna may be scarcer with increasing distance from an inflow, and the recent findings in shallow water of Windermere indicate that Humphries may have been sampling in the least productive part. Numbers in different lakes cannot be compared reliably until the life histories of the various species present are known. Another source of error, to which Jonásson (1958) has drawn attention, is the use of sieves with too large a mesh.

CHAPTER 11

Fish

Anglers and others have been writing about the fish of the Lake District for some two hundred years, without, however, recording any considerable amount of information valuable to the present-day scientist; it is not always clear to which species some of the names used by the earliest writers refer. For those interested only in the scientific aspects, it is sufficient to go back no further than Watson (1925), whose book is particularly valuable as a source of information about man's interference. During the last century there were commercial net fisheries in many of the lakes but they were abandoned because they gradually ceased to be profitable; this may have been another effect of the coming of the railways, which opened up the possibility of obtaining sea fish in reasonably fresh condition. In recent years some of the surviving netting rights have been bought up by persons and by organizations who, in the interests of sport fishery, wished to make certain that they would never be used again. Watson, who held some advanced ideas, would not have supported this step, as he believed that the cessation of netting led to an increase in the number of fish and a resulting diminution in average size. Today it would be rash to express an opinion on an issue of this kind, for recent work has shown that the interaction of factors governing the numbers of fish is sufficiently complex to make the outcome of any measure hard to foretell.

Towards the end of the last century, there were attempts to improve the lakes as places where fish could be caught for sport. Hatcheries were established beside several lakes and there was much transference of fish from one lake to another in the belief, still not dead in fishing circles, that small size was attributable to centuries of inbreeding. These introductions do not seem to have received the notice that is their due from some taxonomists.

After the First World War little was done to improve the fisheries until the Freshwater Biological Association took a hand at the outbreak of the second. It is impossible to say how far this inaction was due to lack of success of earlier attempts, to a general shortage of money, or to the difficulty of protecting the fishing against exploitation by the ever-increasing number of visitors. For many years now there has been little or no restriction on most of the lakes, and anybody who has purchased a river board (now river authority) licence may fish where he pleases. There are laws about the methods which may be used and bailiffs to enforce them.

Trout-fishing with rod and line is popular, though the catches are not good in comparison with, for example, Loch Leven in Scotland and many rivers. Char fishing is an esoteric sport practised mainly by local people, who row slowly down the lake with a rod projecting on either side of their boat. The line is armed with several spinners and hooks, the lowest of which may be as much as 30 m below the surface. When a fish is hooked, the jerk rings a bell at the end of the rod and the fisherman leaves his oars, which are on thole-pins not in rowlocks, and hauls in the line. A spinner at the end of a line towed behind a boat in shallow water is used mainly for pike, though the fisherman using this method may also catch other fish including trout much larger than any that the rod-and-line angler is likely to land. Most anglers sit on the bank with a baited hook and catch perch, though their bag is often a mixed one. It is against the law to leave a baited hook unattended. Considerable numbers of these lakeside anglers come to the lakes from elsewhere for a fishing holiday or for a day's fishing, and it is impossible even to estimate how many fish they remove.

Scientific work on fish started soon after the establishment of the Freshwater Biological Association and K. R. Allen was active until his departure for New Zealand in 1938. Work on fish was not easy in the early years, as there were only eight members on the staff and netting requires a team. Voluntary labour during the war and a much more numerous staff after it made a great increase in the scope of the work possible.

The plan of this chapter is to give a list of the species recorded, to describe briefly the natural history of each one, to discuss their food, and finally to give an account of an experiment on altering the fish population in Windermere.

If there was some confusion over vernacular names in the early writings, scientists are in no position to point a finger of scorn at the authors, for there has been even more confusion over the Latin names. All our trout are now regarded as belonging to the one species *Salmo trutta* Linn. but at one time a number of species were recognized; Watson (1925) mentions several though in places does imply doubt about the validity of some of them. Regan (1911) recognized two species of *Salvelinus*, one, *S. lonsdalei* Regan, occurring in Haweswater and the other, *S. willughbii* (Günther), occurring in the rest. It was fashionable at one period (e.g. Swynnerton and Worthington 1940, Frost 1946a) to reduce these to subspecies of *Salvelinus alpinus*, but now there has been a reversion to Regan's species (e.g. Frost 1965). *S. lonsdalei* has never been sampled and examined regularly and until this has been done, its status must be regarded as uncertain.

Coregonus is still in a state of confusion and the following is a summary of the situation:

English name	Lakes in which found	Regan 1908	Regan 1911	Svärdson 1957	Dottrens 1959
Skelly or Schelly	Ullswater, Haweswater	*clupeoides stigmaticus* Regan	*stigmaticus* Regan	*oxyrhynchus* (Linn.)	*wartmanni* (Bloch)

Vendace	Derwentwater	*vendasius*	*gracilior*	*albula*
	Bassenthwaite	*gracilior*	Regan	(Linn.)
		Regan		

Dottrens expresses the opinion that further ecological and biometric studies are necessary to determine exactly the status of the British populations. Until this has been done, the English names are likely to lead to greater clarity than the Latin ones.

The remaining species of fish, being less variable and, it would seem, less liable to isolation, have not produced similar problems.

It is not easy to discover whether a fish is present or absent in a lake or even to determine its numerical status. Moreover fishermen, on whose records a list must depend to some extent, tend to ignore the small fish. Accordingly the list below is of the species in Windermere, which has been well studied, and the other lakes are compared with it:

Salmo salar Linn.	salmon
Salmo trutta Linn.	trout
Salvelinus willughbii (Günther)	char
Perca fluviatilis Linn.	perch
Esox lucius Linn.	pike
Anguilla anguilla (Linn.)	eel
Phoxinus phoxinus (Linn.)	minnow
Gasterosteus aculeatus (Linn.)	stickleback
Cottus gobio (Linn.)	bullhead
Scardinius erythrophthalmus (Linn.)	rudd
Leuciscus rutilus (Linn.)	roach
Tinca tinca (Linn.)	tench

These twelve species fall evenly into four groups which might be designated: game fish, common coarse fish, small fish, and rare coarse fish.

An occasional salmon passes through Windermere and other lakes too, but numbers are very small. Only two tench have been caught. The stone-loach, *Nemacheilus barbatulus* (Linn.), is a species usually found in running water, but Smyly (1955) records that it does extend a little way into Esthwaite. McCormack (1965) records the stone-loach in Ullswater, without, however, stating whether it is a true lake species rather than a delta fish. She mentions also the lampreys, *Lampetra planeri* Block and L. *fluviatilis* Linn.

The remainder may be considered in terms of the lake series. Pearsall (1921) pointed out that trout dominate the least productive lakes, in which char also occur, and that pike and perch are absent from Wastwater though there were said to be a few perch in Ennerdale. Absence is hard to prove, but the difference between absence and scarcity is less important than the difference between scarcity and abundance, which can be established more easily. How far scarcity is due to poor food supply and how far to lack of suitable breeding grounds is not known. Perch are present in all the other lakes and very abundant in the more productive ones. The same is true of the pike except that it has not been recorded from Ullswater and Haweswater. The eel is probably ubiquitous.

The trout persists alongside these coarse fish in all the lakes but char are believed not to occur in Derwentwater, Bassenthwaite, Ullswater and Esthwaite.

The 'rare coarse fish' have possibly been introduced recently into Windermere either deliberately or accidentally by anglers. As far as is known they remain rare, but from Esthwaite come reports, unsupported as yet by scientific study, that roach and rudd are numerous.

Little is known about the occurrence of the 'small fish', which are ignored by anglers. The abundance of sticklebacks in Ennerdale was established by Mr G. Thompson who caught them in traps set in deep water for *Mysis*.

Four lakes, as already noted, contain *Coregonus*, but how far this is a historical accident and how far a reflexion of the conditions in them cannot be said. Derwentwater and Bassenthwaite are broad and shallow lakes compared with the rest, and it is noteworthy that these are two in which there are no char. Ullswater and Haweswater, on the other hand, have the narrow, deep basins more typical of the Lake District lakes. These four lakes lie in valleys facing north and north-west, but it is difficult to explain present distribution in terms of conditions obtaining at the time of the original invasion without running into difficulties over absence in Buttermere, Crummock and Loweswater. Even if these were not reached by the original immigration, it is difficult to see why, in later years, they have not been reached by a short journey down the R. Derwent from Bassenthwaite and up the R. Cocker to Crummock Water (fig. 4). In Haweswater char and skelly both appear to be abundant, but in Ullswater the char is believed to be extinct though there are reliable records that it occurred there formerly. Nilsson (1963) states that *Salvelinus* sometimes disappears from a lake into which *Coregonus* has been introduced.

If these two species be treated together the lakes can be arranged as follows according to the abundant species:

En, Wa	the rest (except Es)	Es
trout	trout	trout
char	char/*Coregonus*	pike
	pike	perch
	perch	rudd
		roach

There are anomalies, however, notably the absence of pike from Ullswater and Haweswater.

NATURAL HISTORY

There is no record of the trout breeding in any Lake District lake, but it would be rash to state categorically that it does not cut a redd and lay eggs in the few places where the substratum is suitable, as this is known to occur elsewhere (Frost and Brown 1967, p. 70). Most, if not all, mature Windermere fish run up streams to breed and the young return when they reach a certain size. According to Allen (1938a) this is attained in one year by 12 per cent, in two years by 75, and in three years by 12 per cent. Information about the rate of growth has been summarized by Swynnerton and Worthington (1939) and is presented here in fig. 56. Frost (1956) recorded more rapid growth in Haweswater after the level had been raised. Allen (1938a) found that 75 per cent of the trout make most of their growth in summer, 25 per cent in winter. He postulates that this may be a misinterpretation of the true course of events due to the failure of some fish to lay down any annuli on the scales in winter. At this time, therefore, there are broad annuli at the edge of the scale which make it appear that the fish is growing. A more likely explanation, supported by data from marked and recaptured fish, is that rapid and slow growth alternate regularly but not at exactly the same time in each specimen; some start and stop early in the year, others late.

Allen caught in a seine net three times as many trout in winter as in summer, which he takes to indicate a partial withdrawal to deeper water during the warmer part of the year.

Fig. 56 compares the growth in Windermere with that in two other Lake District lakes and in two Irish lakes. Reference to Appendix IV in Frost and Brown (1967) shows that the two Irish lakes are near the extremes, though there is one where growth is better than in L. Derg and several where it is slightly worse than in L. Atorick. Wastwater resembles Haweswater in that the growth of trout in it is very poor. Growth in Windermere is shown to be moderately good by this comparison.

The eggs of the char take between two and two and a half months to hatch. Little is known about the fry. The adult fish, as far as can be ascertained, spend most of their time in deep water though occasionally they come to the layers near the surface. Some of the Windermere char spawn in shallow water in the autumn, others in deep water in the spring, and there appears to be no mingling of the two. This is the feature of the species that has attracted most scientific study (Frost 1965). The char spawns on a substratum of gravel or small stones, and buries its eggs in a redd, as do salmon and trout. Autumn spawning lasts for some ten weeks in October, November and December, with the greatest activity in November. The spawning grounds are either in the lake at a depth of from 1 to 3 m or in a large pool in the River Brathay. The spring spawning lasts for some fifteen weeks in January, February, March and April, with greatest activity in February and early March. Both these periods are equidistant from the winter solstice and are separated by a period of a few weeks when no eggs are being laid. A suitable substratum in deep water must be of limited occurrence because most of the lake bed below 3 or 4 m is covered with

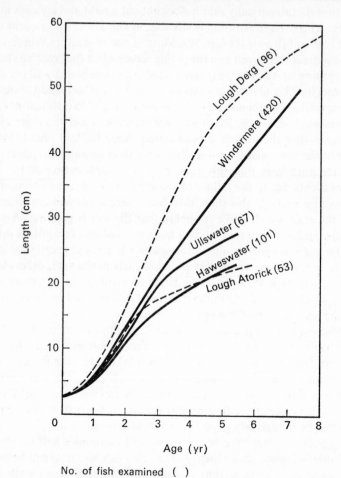

No. of fish examined ( )

56. The average growth rates of brown trout in lakes (Swynnerton, G. H. and Worthington, E. B. (1939), *Salm. Trout Mag.* **97**).

mud. One has been investigated by divers who report a tongue of stones and gravel 14–16 m wide extending down to a depth of 28 m. It is off the mouth of the Holbeck, which, presumably is either building a delta sufficiently fast to prevent the coarse being smothered under fine material, or flowing sufficiently fast down the floor of the lake to prevent silt accumulating. The fish lay their eggs in the lower half of this tongue. The position of other spring spawning grounds is known from local information and from the occasional capture of ripe char in gill-nets set for pike.

The number of gill-rakers on the lower part of the first gill arch is:

autumn spawners in the lake   11–16 mean 13·3
spring spawners   13–17   „   15·1

There is also a small difference in the scales. Freezing of the sperm has made artificial crossing possible and the hybrids produced sperm and ova. An epidemic in the rearing ponds brought further investigation to an end.

No less than 584 out of 2066 fish tagged have been recaptured, all on the spawning bed where they were first caught and marked. Planting experiments in various becks where no natural run was known has been followed by the appearance of spawning fish, but the difficulty of making certain that there was no natural run makes this evidence less convincing. It seems highly probable, however, that the two groups are kept completely apart by their different times and places of spawning. Frost (1965) discusses whether this is an example of sympatric speciation or whether there has been a geographical barrier, but the data from Windermere contribute nothing towards a solution of this controversy and further discussion would, therefore, be out of place here.

Some 20 per cent of all fish spawn twice, more males than females returning, and approximately 1 per cent and 0·1 per cent come back a third and fourth time.

Spawning in spring in deep water is known to take place in Coniston, Haweswater and Buttermere, and the fish of Ennerdale run up the River Liza to spawn in autumn. The char in Crummock spawn in spring but it is not known where.

Knowledge about *Coregonus* is meagre, and it illustrates the pitfalls lying in the path of anyone who bases statements on 'local information'. I have before me a sheet of information culled from my colleagues and based on the written records, on personal experience, and on communications from fishermen and other local inhabitants. The information against the Derwentwater vendace is 'one washed up' and against the Bassenthwaite vendace 'one washed up in 1940', the implication being extreme rarity. However, when Maitland (1966) visited the Lake District with experience of how to catch this species, he found flourishing populations in both lakes.

Bagenal (1966) records that the skelly spawns in 2–6 m of water from early January to mid-February and lays its eggs on the bottom with little protection. In Loch Lomond, Slack (1957) observed that 65 per cent of eggs laid in this way are eaten by larvae of *Phryganea*. The following lengths at different ages are quoted by Bagenal (1966):

Length in cm	20	25	30	33
Age in years	2	3	4	5

The largest specimen was 40 cm long and weighed 680 g.

In January 1966 a number of skelly were cast ashore during a gale, a phenomenon that has been recorded before. Bagenal suggests that, having come from deep water to spawn, they were still imperfectly adjusted to life in shallow water and were overcome by the turbulence caused by the strong wind.

Ellison (1966) has brought together old published records and local lore. Formerly vast shoals used to assemble in shallow water during the harvest months, when they were netted in great quantity. Dorothy Wordsworth's journal contains an account of a large haul in 1805. These big shoals have not been observed recently.

Perch are found at depths of between 18 and 27 m in Windermere during the winter and they move into shallow water to spawn during May and early June. The eggs are laid on rooted plants especially *Elodea* (Allen 1935). More information about them was obtained during the course of trapping experiments which are described later.

Pike spawn in reed-beds and later move into deeper water. Frost and Kipling (1961) have studied the growth of this fish (fig. 57).

Eels, most of them female, enter Windermere in fair numbers and stay there for from nine to nineteen years (average 12·27) before setting out on the return journey to the sea, generally during a time of flood in the autumn. Their length just before migration ranges from 49 to 95 cm and their weight from 210 to 2040 g. The males leave after between seven and twelve years (average 9) at a mean length of 40 cm and a mean weight of 112 g (Frost 1945).

Minnows occur in shoals over stony substrata in Windermere, at a depth of less than 1 m in summer but at about 2 m in winter, when they are frequently found under the stones. Spawning takes place in May, June and July on the lake bottom and also in streams, up which the fish migrate. Some are mature after one year, and three, possibly four, year-classes can be distinguished (Frost 1943).

The bullhead lives under stones. In April the male makes a nest and the female comes to him to mate and lay eggs. She then departs leaving the male to brood over the eggs and keep a current of water flowing over them (Smyly 1957).

No study has been made of the stickleback.

The food of these species has been studied by the authors quoted and several other papers have been devoted exclusively to this topic.

Char stomachs generally contain little but zooplankton, though the spinners on which fishermen catch them are far larger than any plankton animal. In spring and early summer chironomid larvae and pupae, and in winter char eggs, are taken in moderate quantity (Frost 1946a). Skellies have a similar diet, but the species found in them suggest that they feed less in the open water and more in weed-beds. However, as the only information available is that derived from a sample of ten fish examined by

Swynnerton and Worthington (1940), a definite statement is premature. All that can be stated with confidence is that they hardly ever take anything offered by fishermen.

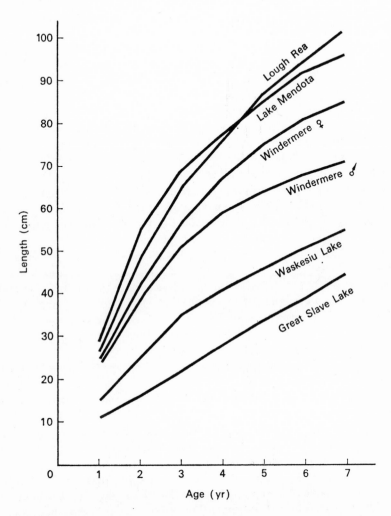

57. Growth of pike from various waters (Frost, W. E. and Kipling C. (1967), *J. Anim. Ecol.* **36**).

The other specialist is the pike which, after a short early period subsisting first on plankton and then on benthic invertebrates, becomes exclusively piscivorous. Perch are eaten throughout the year and are the main item of diet from March to October. In November and December char are the main prey, giving place to trout in January and February

(Frost 1954). Allen (1939) found empty stomachs at all times of year, but more in winter than in summer.

The remainder, trout, perch, eels, minnows and bullheads all feed extensively on the invertebrates inhabiting the bottom in shallow water, though the first two change to fish when they are large. Trout feed mainly on the permanent fauna, that is *Gammarus, Asellus* and molluscs, from October to February with a period in November when char eggs are the main food. In March chironomids are important, followed by nymphs of *Nemoura* in April and larvae of *Leptocerus* in May and June. For the rest of the summer the main prey is found on the surface during the daytime but there is a change to bottom-feeding at night. A test applied during the period when the fish were taking aquatic food showed that 65 per cent of 180 individuals were selecting one food organism, 6 per cent were selecting two and the remaining 29 per cent were taking prey unselectively. At a length of about 40 cm trout take to a diet of fish (Allen 1938a).

Unfortunately little is known about the bottom fauna at the time of Allen's work, which makes speculation about the accuracy of some of his statements inevitable. The names quoted in the preceding paragraph are the only ones to be found in both the table and the text of Allen's paper, and it is difficult to know how to interpret them. There are two species of *Nemoura* in Windermere, *N. avicularis* on the stony substratum and *N. cinerea* in reed-beds, but their numbers are smaller than those of several other species of Plecoptera. It is possible that Allen used the name *Nemoura* to include more species than it does today. If, as is likely, *Leptocerus* was the name given to any caddis in a stone case, it covered several other genera, some in different families.

Macan, McCormack and Maudsley (1966–7) studied the food of trout in a small tarn and make certain observations that are relevant to the present theme. They too found that trout feed extensively on animals in the surface film during the summer, but not regularly. On some days, generally wet and windy ones, most of the food was aquatic organisms. It is, therefore, suggested that the course of events in Windermere may not be quite as regular as Allen makes it out to be.

Allen writes: 'The nature of the food of the fish is controlled first by what is available and secondly by the behaviour of the fish.' The last part of the sentence is incontrovertible; if trout behaved like carp and grubbed in the bottom, or like perch and did not rise to organisms at the surface, the nature of its food would be different. The term 'available' merits further examination. Availability depends first on the behaviour of the organism. If an animal lives in the mud it is safe from predation by trout. The availability may change suddenly if, for example, an insect has to leave the mud to emerge. Chironomids are particularly vulnerable as they make what may be a long journey through the water to the surface. The duration of availability is of considerable practical importance. Macan

(1965a) discusses the availability of water bugs to trout. Frost and Macan (1948) had shown that these animals are rarely numerous inside fish and are not often found there at all, which they attributed to unpalatability. Macan's later conclusion was that, on the contrary, it is because the animals are readily eaten and easily caught, in other words easily available, that they are not prominent in the diet of fish. Any that do occur in a fish's feeding zone are eaten soon and, as the breeding season is short and the development quick, they cannot be replaced. An organism with this type of life history can never be an important item of food to fish except perhaps for a short time in places where its living space is large and the fish's living space is small. Generally the reverse is found. The habitat of such an organism must obviously be a place that gives protection against predation.

The organisms important as a source of food for fish are those which, though protected, are not too securely protected against predation; in other words their availability is only moderate. They must also possess some mechanism whereby losses are quickly made good. This is achieved in various ways. *Gammarus*, for example, breeds from January to October. Some insects have a short breeding season but the hatching of the eggs is spread over a long period. Others maintain a reserve of small specimens, suppressing their growth by keeping them out of the good feeding places or by secreting some inhibitor. These small specimens, which are ignored by trout, are available to fill each gap caused by the death of a large specimen. Once a favourable niche is secured growth is rapid, and this is offered as an explanation of how large specimens remain little changed in numbers whether predation be absent or intense (refs. in Macan, McCormack and Maudsley (1966/67).

Allen (1935) also studied the food of perch. In summer when the fish are in shallow water, the smallest, those under 16·5 cm long, are eating plankton. As length increases up to about 19 cm, more and more bottom-living invertebrates are eaten, and above that size the diet is mainly fish. Minnows are the main item of prey and sticklebacks and small perch are also eaten. In winter plankton is eaten more by fish of all sizes. The invertebrate prey mentioned by Allen comprises chironomid larvae and pupae and *Gammarus* at all seasons, *Chaoborus* larvae and *Asellus* only in winter and *Ephemera* and *Sialis* only in summer. The identification of the last two genera cannot have provided any difficulty and the record can be relied on. Its importance is discussed shortly.

McCormack (in press) has examined 1128 perch caught during the years 1963 to 1967. *Asellus* is now a more important item of diet than it was at the time of Allen's work. The most striking difference is that the change of diet with size is much less marked in the post-trapping era than it was before; plankton is eaten more by fish of all sizes and the largest are less exclusively piscivorous than formerly. McCormack was on the look-out for evidence of sufficient predation by older fish on the new generation to annihilate it, as postulated by Alm, and did not find it.

Asellus is probably more abundant today than it was and this is probably the reason why more are eaten. There is no explanation of the other changes. It is not possible to make close comparison with what the perch is eating and what occurs on the feeding grounds as insects are identified to orders only.

Cladocera, Copepoda, small benthic invertebrates, and some algae were found in minnows by Frost (1943). Bullheads feed on benthic organisms of all kinds and, in an aquarium, take only moving forms. Smyly (1957) also observed that they feed at night, which, together with their habit of resting beneath stones, is no doubt the reason why so few are eaten by other fish. Of just over 2000 organisms found in 106 eels from Windermere, 1504, which is 75 per cent, were Mollusca. Thirty per cent were *Valvata piscinalis*, 22 per cent were small bivalves, 18 per cent were *Limnaea pereger* and the remaining 5 per cent were made up of *Ancylus fluviatilis*, species of *Planorbis* and unidentified specimens. This indicates that the fish are foraging mainly among rooted plants. Thirty-eight *Sialis* larvae and twenty-two nymphs of *Ephemera danica* suggest that the eel can also find food in the bottom mud. The food of forty-two fish taken in Cunsey Beck, which flows out of Esthwaite, was different, 113 out of 145 organisms eaten being either polycentropid larvae or *Ephemerella ignita* nymphs. In contrast again thirty-two eels from the River Leven had been feeding mainly on chironomids. Other fish had been eaten only rarely by eels from all three places (Frost 1946b).

The char and the pike and possibly *Coregonus* occupy unique positions in any scheme of food relationships. On the other hand trout, perch, eel, bullheads, minnows and probably also sticklebacks, which have not been investigated, all eat the same kind of food at some time of their lives. That reduction in the number of fish has led to increased size among the survivors indicates that total biomass is limited by the food supply. Is it preferable that six species should contribute to this biomass or would it be advantageous to reduce the number, if this could be done? The answer to that question depends on who gives it. The angler frequently strives to destroy all fish except the one which is the object of his sport. In Windermere there has been an increase in the number of trout; Frost (1954) makes this claim on the grounds that the average catch in standard hauls with a seine net was nearly five times as great in the years 1945–1952 as it was in the four that preceded them. Unfortunately it is not possible to distinguish how far this has been caused by the reduction of the competitor, perch, and how far to the reduction of the predator, pike. As far as can be seen from the facts available at the moment, further improvement would result from the reduction of the numbers of bullheads and eels, or indeed from their eradication, if this were possible. The status of the minnow is un-

certain. Its main diet of small crustacea is probably being replenished all
the time as the open water and its plankton is blown to the edge of the lake.
The volume of water is so large relative to that inhabited by the fish, and
plankton reproduces so fast, that exploitation is likely to be incomplete.
The minnow therefore has no disadvantage as a competitor, and may have
the advantage of being the intermediate stage whereby plankton organisms
are made available to fish too large to eat them directly. Allen (1938a),
however, found that by the time trout take to a diet consisting mainly of
fish, they have reached a size at which they generally eat prey larger than
minnows. On the other hand, perch eat many (Allen 1935) and anybody
concerned with the success of that fish would certainly encourage the
minnow.

Anyone interested in total production would be likely to advocate
additions to rather than subtractions from the existing species, for if
suitable kinds exist or could be found, production in Windermere would
benefit from the introduction of a herbivore, or of a species that could grub
in the bottom like a carp. The practical question is the extent to which
existing species are exploiting the resources in the same way. The trout is a
cold-water species that breeds in winter, the perch a warm-water species
that breeds in summer, but there is no great difference in their feeding
activity during a year. Allen (1938a) concludes somewhat tentatively that
trout eat more in summer than in winter. Macan, McCormack and
Maudsley (1966-7) found that, though fish with full stomachs were taken
from a moorland fishpond at all times of year, they contributed a higher
proportion of the total catch in the months of May and June than at other
times. The fish were more active at that time to judge by the ease with
which they were captured in gill nets. This confirmed the findings of
Swift, who studied the growth rate in the laboratory (1961) and the movement of trout confined in a cage in the lake (1962, 1964).

Both perch (Allen 1935) and the bullhead (Smyly 1957) feed more
actively in summer than in winter. Eels have not been sampled throughout
the year. The work carried out so far has, therefore, produced no evidence
that different species, their different temperature requirements notwithstanding, exploit the available resources at different times. It does produce
evidence that the exploitation is not the same. Only trout eat the food at the
surface. Perch, in contrast, appear to be able to obtain food from the mud,
since both *Sialis* and *Ephemera* feature fairly prominently in the list of
what they eat. That eels, with their elongate form and ability to proceed
along a firm surface, have feeding habits unlike the two species mentioned
cannot be doubted.

The significance of different food habits can be judged only when something is known about the behaviour when alone of species that are generally
found together. Swedish workers, who have made extensive and important
contributions to the study of fish food, have investigated this problem

also, and Nilsson (1965) has produced a short review of it. When char and trout occur together in north Swedish lakes, trout inhabit the shallow water and feed on the bottom animals, turning to surface food late in the summer. Char feed on bottom organisms in the winter, on surface food in June, and on plankton for the rest of the summer. When trout are absent, char occur in the shallow water as well, and there they feed on the bottom invertebrates as trout do where they are present. In one lake there was a group of large char following the way of life characteristic of trout when both species are present, and a group of smaller char occupying the region to which the species is confined in the presence of trout.

There are no comparable observations for other fish and, as Nilsson points out, it cannot be assumed that what he observed in north Sweden would happen under all conditions. It seems unlikely that the interaction between trout and perch would be as great as that between trout and char, which, Nilsson notes, have almost identical food preferences when free to exercise them without constraint. It may be concluded that the existing diversity of species does lead to greater total annual production of fish flesh in Windermere.

The first calculation of the total population of any species was that of Allen (1938a) who based a figure for the number of trout in Windermere on the number of previously marked fish recaptured in a seine net and an estimate of the number small enough to pass through the net. He reckoned that there were some 12,000 trout in the littoral region of Windermere. In a popular account of the same work (1938b), he includes an estimate of numbers beyond the reach of his seine net, and arrives at a total population of between 15,000 and 20,000. This he believes to be smaller than formerly and writes: 'It is generally agreed among those whose experience is sufficiently extensive that there has been a decided deterioration in the trout fishing in Windermere during the last thirty years. This deterioration is almost certainly due to a decrease in the number of trout in the lake.' Perusal of what has been written about fish in the English Lake District has left the present writer sceptical about the value of such statements unsupported by any kind of record made at the time. Even if it is true that fishermen catch fewer fish, the explanation could be that, as a result of enrichment by sewage, there is now more aquatic food available and fish rise less often to what the fisherman offers them. Nilsson (1965) has recorded that trout feed more at the surface if the underwater food supply is reduced. Allen's explanation is that fewer fish now survive in the becks because more succumb to floods, more frequent than formerly owing to improved drainage, to sheep dip or to toxic substances from the roads. This is speculation permissible to those interested in improved fishing, but it is not a statement supported by scientific measurement.

Numbers of perch have dropped in the last thirty years, and this statement is made with no fear of contradiction because the drop was caused by trapping and proper records were kept. The work started early in the war with the immediate object of adding to the nation's food supply and the ultimate object of producing fish that were larger because they were fewer. The perch were caught in wire traps with a conical inverted entrance, and they went into them in search of a place to spawn, or out of curiosity, when on their way from the deep water to spawn in the shallow. The traps were generally set in late April and lifted once or twice a week according to the numbers captured. The period when catches were large enough to make lifting worth while was generally six or seven weeks, and the start of it varied a little according to the temperature of the water. A line of pilot traps down the slope of the lake floor indicated the depth at which fish were densest and the others were set at this depth. The main concentration of fish moved slowly into shallow water as the season progressed. Experimental trials were held in 1940, 1941 was a year of intensive effort largely confined to the North Basin and this was continued and extended to both basins until the end of the 1947 season. Six to seven hundred traps were operated, most of them by the voluntary labour of local residents. The catch was canned and put on the market.

In 1948 and subsequent years a few traps were put down to obtain a sample of fish, and the cooperative effort was discontinued.

In 1942, the first year when the whole lake was fished, a little over 3000 Kg of perch was removed. About half that weight was taken out in 1943 and in 1944 the catch was again about 50 per cent of that of the year before. Thereafter the catch was more constant each year (Worthington 1950). Le Cren (1958, 1959) has published the main account of the results up to 1955 followed by a short version. His graph (fig. 58) shows the relative population density reduced to a level of about 3 per cent, with biomass, at 8 per cent, higher owing to improved growth by survivors. His 'relative population density' is in fact based on catch per unit effort and he regards it as a direct measure of the actual population. Macan and Worthington (1951, p. 208) express doubt that the relationship is as simple as this. They suggest that, when the population is dense, the pressure to expand into any area where population has been reduced, by trapping for example, must be great. As population density decreases this pressure to expand decreases and there must come a level when it ceases. The proportion caught is not, therefore, constant but decreases regularly as total numbers fall. The present estimate of a population now 3 per cent of what it was formerly is too low according to this argument, but the error is not likely to be great, and there is no doubt that the trapping has reduced the number of perch drastically.

While the question of numbers is still unsettled, investigations have centred round growth and recovery, in both of which fields there have been surprises. Most growth takes place between June and September and the rate varies from year to year. Careful analysis has shown a fairly good

N

correlation with number of degree hours above 14°C. In order to compare the years, the rate each year has been corrected to show what it would have been had temperature not varied (fig. 59). Growth during the first two years was similar for both sexes throughout the period of observation, though there were always a few specimens that grew distinctly faster than the rest. After the second year the annual rate of growth increased with each successive year-class. The unexpected finding was that improved conditions for growth, presumably related to the reduction in population, did not

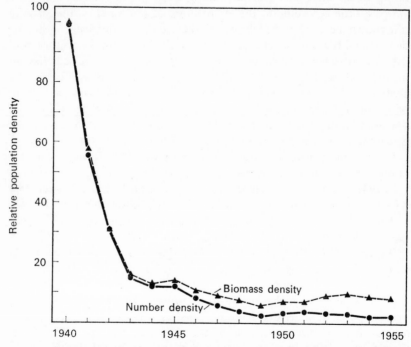

58. Changes in population density of perch in Windermere. The geometric means of the catches per trap made in successive pairs of years from 1940–1941 to 1955–1956 expressed as a percentage of the 1940–1941 value. The mean for each pair is plotted against the first year as indicating the population density in the summer growing season immediately following the trapping season (Le Cren, E. D. (1958), *J. Anim. Ecol.* **27**).

affect fish more than two years old. They continued to grow at a steady rate while later year-classes were growing faster.

Before 1941 the annual weight increment was about 5 g. In 1941 fish of the 1938 and 1939 year-classes increased in weight by about 10 g. Females of the 1952 year-class added some 52 g to their weight each year and males of the 1953 year-class some 37 g. The average weight of perch caught by fishermen rose from about 65 g before 1941 to about 170 g in 1957.

The unexpected feature of fig. 58 is that, eight years after intensive fishing ceased, there is no sign of the recovery of which an animal laying as many eggs as the perch is obviously capable within a short time. The success of the year-classes varies considerably, and in 1952, for example, three-year-olds were thirty times as numerous as in 1951 and 1953 (Le

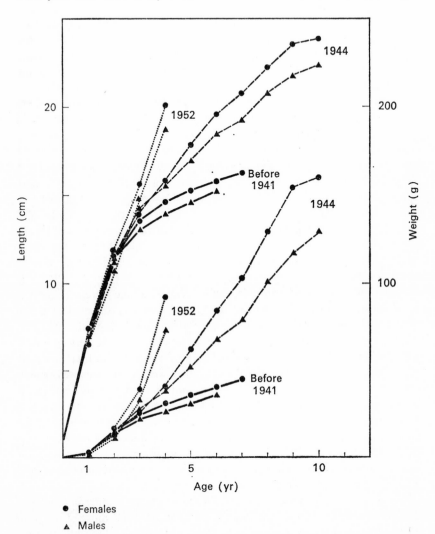

59. Length-for-age and weight-for-age growth curves for male and female fish of the combined year-classes before 1941 and the 1944 and 1952 year-classes. The data have been adjusted for year-to-year temperature variations (Le Cren, E. D. (1958), *J. Anim. Ecol.* **27**).

Cren 1955). The four-year-olds of the most successful year-class were one hundred times as numerous as those of the least successful (Le Cren 1961). These numbers bear no relation to the number of eggs laid, but do appear to be related to climate, a warm summer favouring survival. That a year-class is strong in all the lakes if it is successful in any is evidence that climate is the factor involved, for it is difficult to see how any other can be common to a number of dissimilar lakes, but the nature of the correlation remains unknown.

These good year-classes have, however, led to no more than temporary increases in the total population, and, in order to examine the explanation of this which has been put forward, it will be necessary to describe measures taken against the pike. The success in reducing the number of perch and the consequent reduction of pressure on their food supply, which is shared by the trout, might have been of no benefit to trout if pike ate more of them because perch were scarcer. Accordingly, in 1944 and subsequent years, gill-nets were set for pike during the winter. It was necessary to use nets with a mesh large enough to allow most trout to pass through and these do not catch pike in their first or second year, and only a small proportion of those in their third year. Older fish are almost all too large to pass through the net. Of the catch of females in 1944, the highest proportion was aged four but the numbers, in the next five year-classes, were similar and only slightly lower. The oldest fish caught during the early years was 17 and some 20 per cent of the population was nine years old or more. The effect of the netting was a drastic reduction in the number of older fish, but no reduction on the total population. This did fall at first and by 1947 was only half what it had been in 1944, but by 1950 it had recovered. There was another decline in the first half of the fifties but it was followed by a rise which took the population in 1960 and 1961 to a level above that of 1944. Survival of the young appears to be better as a result of the netting, probably owing to a reduction of cannibalism by large fish, and there are now more three-year-old pike in the lake than there were when measures against the fish started (Frost and Kipling 1967 a and b). Predation by large perch is now greater than it was before 1941 because the quicker growth produces specimens of fish-eating size much sooner. Le Cren (1958) reckons that the biomass of fish-eating perch had regained the 1940 level by 1949 and was $2\frac{1}{2}$ times greater by 1955. This increased cannibalism together with increased predation by small pike is thought to be maintaining the total perch population at a steady low level.

Ullswater, in which there are no pike, differs from Windermere in that a successful year-class dominates the population much more. Fish were trapped in 1941 but not again until 1953, after which observations were made each year. The catches in 1941 and 1953 were similar and the average weight of a perch was 55 g. In 1954 and 1955 the total number caught, the average catch per trap and the total weight caught were all falling, but the average weight was rising. It reached 74·4 g in 1955. These figures represent almost entirely the decline of the 1949 year-class. In 1957 more fish were caught but their average size was less and this was due to the entry of the moderately successful 1955 year-class (McCormack 1965). These results are similar to those of Alm (1952) in Swedish tarns, except that Alm found a more complete dominance by one year-class for a much longer time. He attributes it to cannibalism, but Le Cren (1965), having in mind the adverse effect of a poor English summer on survival of a year-class in a lake, suggests that in a tarn further north breeding may be completely fruitless except in unusually favourable years.

CHAPTER 12

History of the Lakes revealed by their Sediments

Started just before the war on the initiative of Dr C. H. Mortimer, and continued ever since, the examination of the muds flooring the lakes to see what information about the past lay concealed in them was a logical development of Pearsall's ideas. Dr W. Pennington (Mrs T. G. Tutin) has been the main worker throughout the period, and a number of others have entered the field for short spells.

The problems of obtaining samples were solved first by B. M. Jenkin and later by F. J. H. Mackereth, whose inventions are described in a later chapter.

The cores proved to be stratified in various ways. There were physical differences immediately evident to the naked eye; samples taken from different parts were found to vary in chemical composition; and the remains of organisms showed that the abundance of various species had changed with the passage of time. By far the most complete record is that of the flowering plants in the drainage area, for pollen grains are preserved in the sediments and many of them can be assigned to a species with certainty. The record of aquatic organisms is more fragmentary because so many decompose leaving no trace. Diatom frustules abound in the cores and most can be identified, but other algae have disappeared and animal remains consist mainly of parts of Cladocera and heads of chironomids.

Assigning dates to the various levels of the cores was the first problem. In the early days Pennington (Tutin 1955) investigated present-day rates of mineral sedimentation, hoping that this had been constant over the years, but Mackereth (1966) showed that the rate of sedimentation is a measure of the rate of erosion, which has not been constant. The work was contemporary with research all over the northern hemisphere devoted to analysing the pollen content of peats and lake sediments at different depths and constructing therefrom a history of the vegetation since the Ice Age, and correlation of particular associations of pollen in the cores with associations discovered elsewhere has been the main method of dating. The age of some of the other pollen zones has been ascertained by means of the C^{14} technique, but facilities for this being limited, it has been applied to few samples from lakes up to the present. The work gradually became more closely related to pollen studies than to limnology, but events in a lake are influenced by events in the drainage basin and are therefore relevant to a study of fresh waters.

The first publication, on the diatoms of Windermere (Pennington 1943), was followed by the first attempt at dating, in which Pearsall and Pennington (1947) related changes in the cores to known historical and prehistorical events. The pollen from Windermere was described the same year by Pennington (1947), and that of Esthwaite by Franks and Pennington (1961). Pennington (1964, 1965) then turned her attention to six upland tarns, believing, correctly, that they would yield information that was not available in the sediments of the main lakes. Round (1957a and 1961c) investigated the diatoms in cores from Kentmere and from Esthwaite. The main zoological study is that of Goulden (1964), and the chemical studies were published by Mackereth (1965, 1966).

In the account which follows, the pollen studies are described first, as they provide means by which depth and age can be related. Mackereth's closely argued deductions from his chemical analysis are taken next, because, although he is not always able to decide which of several possibilities is correct, all the entities in his field are present. Lastly the inevitably incomplete data about aquatic organisms are presented.

In the littoral region, a mixture of clay and stones has been found to overlie the rock of the lake basin, but no sampler has penetrated this layer in deep water. On the basal stony clay lies laminated clay, greyish in Esthwaite but more pink in Windermere. The next sediment, referred to as silty detritus mud, is generally in the form of a conspicuous grey layer, though in some cores from the North Basin of Windermere it is thin and hard to make out. On this grey layer lies more clay, which in Esthwaite is stony and unlaminated but which in Windermere is laminated. On top of this is the thickest of the explored layers, a uniform brown mud that extends almost to the surface. It is capped by black fluid surface ooze (figs. 65 and 67).

The lowest clays were deposited during the Ice Age. When the glacier snout lay over the present lake basin, or not far above it, morainic material of all sizes fell to the bottom of the lake, but, when the snout had retreated higher up the valley, only fine particles were carried in. The largest settled at once. In winter the inflow froze and there was no further supply, but very fine particles, afloat since the summer, continued to settle, which produced the laminated structure evident today. The silty detritus mud represents an amelioration of glacial conditions and the next band of clay their return. The brown mud accumulated during the post-glacial period.

The climatic amelioration that intervened at the end of the glacial period is known as the Allerød oscillation or interstadial from the place in Denmark where it was first described. Godwin (1961) subjected a sample from this layer in Low Wray Bay, Windermere, to radio-carbon analysis and obtained a date of 9920 ± 120 B.C., which agrees well with dates from elsewhere. The age of any given level in the brown mud has been determined by comparison, as described above. Botanists have divided post-glacial sediments into zones to facilitate comparison.

The history of the vegetation of the Lake District is then as follows (Pennington 1947, Franks and Pennington 1961):

Zone I—the laminated clay, was laid down towards the end of the Ice Age. The soil left bare by the retreating ice was eventually colonized by herbaceous species, and the pollen of a few arctic-alpine shrubs, mainly *Betula nana* and *Salix herbacea*, is found in the earliest layers of the silty detritus mud.

Zone II—Allerød, is the warmer period, in which *Betula pubescens* and *B. pendula* were present, though the persistence of herbs, some of them intolerant of shade indicates that a continuous canopy of trees did not develop. Pollen grains of *Myriophyllum alterniflorum* and *M. spicatum* are frequent, and that of *Nuphar sp., Littorella uniflora* and *Typha sp.* occurs.

Zone III—is the return of the glacial conditions and the reversion of the flora to one tolerant of a severe climate.

Zone IV—Preboreal, a period of transition from an arctic to a temperate climate, is dominated by *Betula*, which was possibly being replaced by *Pinus* towards the end. *Corylus* and *Alnus* are present but scarce.

Zones V and VI—Boreal, a dry period in which the increasing amount of *Corylus* in the *Betula-Pinus* woods is thought to indicate a further increase in temperature. Pollen of *Ulmus* and *Quercus*, also indicators of warmer conditions, becomes more abundant, particularly towards the end of the period, and *Tilia, Hedera* and *Ilex* appear for the first time.

Zone VIIa—Atlantic, a warm wet period, in which the abundance of *Alnus* increases greatly. *Betula* is still common, trees of the mixed-oak forest have increased since the previous zone, whereas *Corylus* and *Pinus* have decreased.

Zone VIIb—Sub-boreal. A drier period in which the peat bogs grew slowly and *Ulmus* began to decline.

Zone VIII—Sub-atlantic, wetter and colder than the preceding, is characterized by a general diminution in the amount of tree-pollen and an increase in the amount derived from herbs.

In the uppermost layers of the mud there is a relative decrease in the amount of *Alnus* and *Betula* and a great increase in the amount of grasses and sedges. *Pinus* increases steadily, and oak gains ground after a decrease.

Pearsall and Pennington (1947) attribute these last changes to man. The *Alnus* decline they associated with the Norse, who cleared and drained the valley floors. *Betula* and *Quercus* declined as the woods were cleared to provide pasture, and grass increased correspondingly. The final increase in *Quercus* is ascribed to planting for amenity. Their approach to the post-Atlantic decrease in woodland is more cautious. The main evidence of it in the core at their disposal was the disappearance of *Pinus*, for which, they point out, two explanations are possible. It could have been due to clearance by man of the upper margins of the woodlands, where, they believed, there was a fringe in which *Pinus* was common. The other possibility is that prolonged leaching had so impoverished the fell tops that bog had developed over much of them, and this bog, spilling over from the points of origin, had overwhelmed the forest. Mackereth (1965) adds a third to these two possibilities, pointing out that a change of climate could have caused a reduction in the number of trees. He notes too that various combinations of these factors could have existed. He stresses that the decrease in the tree cover took place over a wide area and that any explanation must take account of this.

Franks and Pennington (1961) offer no comment on the event, and Pennington (1964, 1965) turned her attention to tarns in the hope of finding more evidence. In the first paper six tarns at various altitudes are discussed and in the second evidence from several more is brought in. In zones I to VI the remains are similar to those found in lakes, and such differences as may be observed are explicable in terms of the different altitudes. For example at the highest, Red Tarn (Langdale) at 1700 feet, clay was still being deposited during the zone IV period, presumably because solifluction was still active at that altitude. The amount of tree pollen relative to that from smaller plants suggests that there was eventually a fairly complete forest cover but it was established later than at lower sites.

Throughout the boreal period there was a heavy deposition of diatom shells but little of material from outside the lake, which is taken to indicate slight erosion owing to the forest cover.

With the coming of the Atlantic period, the organic content of the muds rose, and there were changes in the proportions of some species of diatoms and Cladocera. The degree of erosion did not increase much during this period and the woods evidently remained intact, providing protection against soil erosion even in the wet climate of the period.

With the transition from zone VIIa to zone VIIb there was a decrease in the amount of *Ulmus* pollen, a decrease now generally attributed to the use of elm leaves for fodder at a time when no grassland existed. Thereafter the course of events was different at different sites. Pennington argues that, if climatic change had been paramount, change would have been more nearly uniform and that the diversity was due to the activities of man. This is in accord with conclusions in other parts of Europe.

Evidence from elsewhere puts the date of the elm decline at between 3400 and 3000 B.C. and indicates that it did not take place at the same time everywhere. Radio-carbon dates have also shown that Neolithic man was active in the Lake District as early as this. This date, some 5000 years ago, is in the middle of the post-glacial period.

At most sites *Fraxinus*, the ash, entered the woodland after the elm decline. This is a species that requires plenty of light and therefore

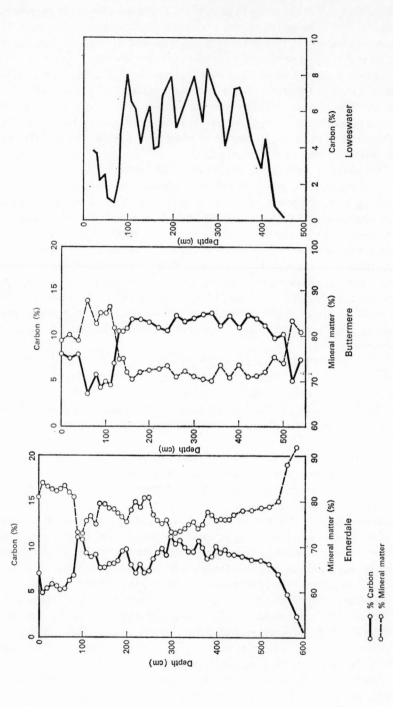

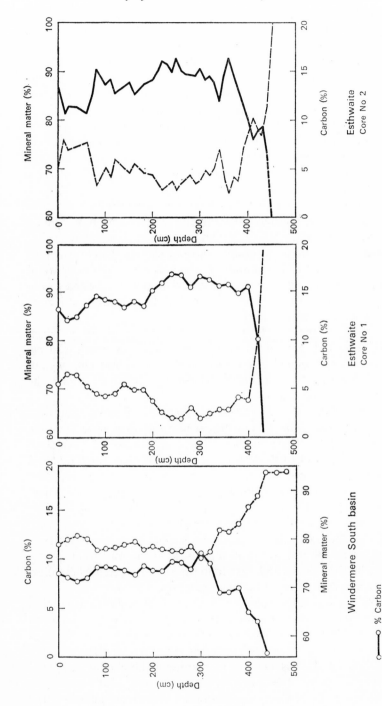

60. Variation of carbon content (percentage of dry material) with depth in the sediments of various valley lakes of the English Lake District (Mackereth, F. J. H. (1966), *Phil. Trans. R. Soc.* **250**).

187

conditions favouring it at the expense of other species are more likely to have resulted from human felling of the woods than from climatic change.

Red Tarn and Blea Tarn both lie near the Langdale axe factory and, at the time of the elm decline, there is a band of mineral silt in the deposit. No such band was seen in any other core, and it is concluded that the forest clearance in this area was so severe or so sudden that erosion was greater than elsewhere. Later there was some regeneration, but elm recovered less than other species. It is suggested that centuries of leaching had reduced the base status of the soil to a level at which *Ulmus* could no longer thrive. The deposits did not accumulate in these two tarns more rapidly after the elm decline than before, which it did in others where human occupation lasted longer.

At Burnmoor Tarn, there is evidence of more regeneration of *Ulmus* than elsewhere and at Thirlmere there is little decline until recent times, owing, it is believed, to the failure of man to penetrate that region.

In contrast, settlement in the drainage area of Devoke Water appears to have been intense and of long duration. The pollen provides evidence of two invasions, the first by a pastoral people who thinned the trees sufficiently to allow grass to grow beneath them, and the second by cultivators who cleared large areas completely.

The removal of the woods increased erosion and leaching, which eventually produced unproductive moorland soils on which *Calluna* flourished. Increased rainfall probably contributed to these changes but Pennington agrees that it could not have been the primary cause since the effect was not observed at all sites.

Mackereth's (1965, 1966) account of chemical differences between levels in the sediments starts from the premise that they came from the land around the lake. The post-glacial sediment is some 4·5 m thick in Esthwaite and 6 m thick in Ennerdale, the reverse of what would obtain if organic matter produced within the lake were the main source. The greater erosion in the steeper more unproductive drainage basin of Ennerdale is his explanation of the thicker deposit in the lake. He also brings forward evidence from recent work that the plants and animals that fall to the mud from the water above decompose leaving little trace. However, some substances—phosphorus was mentioned in an earlier chapter—accumulate in living tissues and, carried to the bottom when the organisms die, may be adsorbed there on various complexes.

Obviously organic matter below the surface of the sediment has undergone prolonged exposure to the forces of decomposition, first on land and then in the top layer of the mud, and would not have persisted if it had not been extremely stable. Mackereth assumes that there is little diffusion within the sediment. He has therefore to interpret his findings in the light of the following variables. The soil is subject to leaching and erosion, the intensity of either of which processes decreases as the other increases. The products of leaching differ according to whether the soil is oxidizing or reducing. Once settled in the water, and while forming the surface layers, the sediment is subjected to both loss and gain, and here too oxygen concentration may determine whether the one or the other takes place. Gain may also be due to the biological causes just mentioned. Composition of the rocks seems to be relatively unimportant though it is mentioned as having an influence on the calcium concentration.

The amount of organic matter in the soil is measured by the carbon content (fig. 60). This rises steeply at the end of the glacial period, then maintains a steady level till about 3000 B.C. after which it drops. There is indication of a further fall about 1200 years ago.

In fig. 61 Mackereth shows the carbon content of three Lake District lakes and Linsley Pond in Connecticut, U.S.A. He comments, 'that the

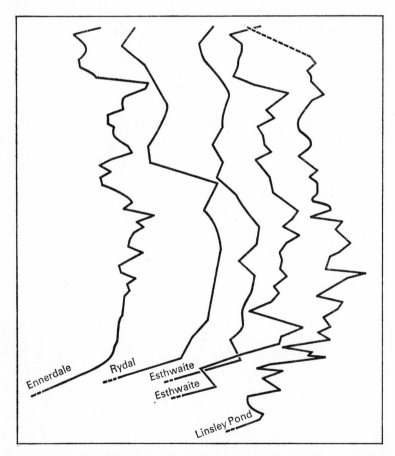

61. Sedimentary carbon profiles from a number of lakes plotted against an approximate time scale. The lower end of the diagram corresponds to an age of roughly 10,000 years (Mackereth, F. J. H. (1966), *Phil. Trans. R. Soc.* **250**).

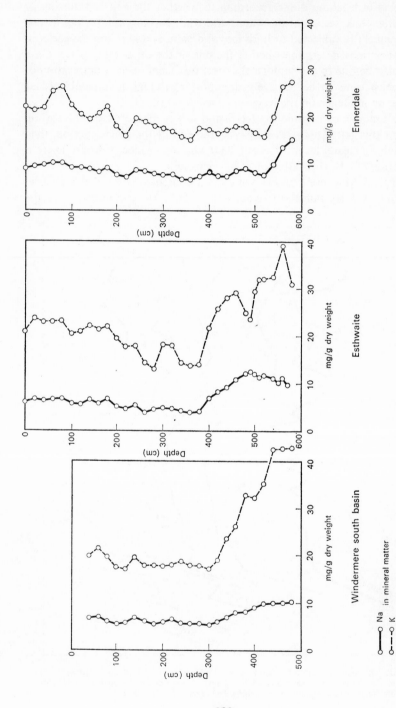

62. Variation with depth in the sediments of three lakes of the sodium and potassium content of the mineral matter (Mackereth, F. J. H. (1966), *Phil. Trans. R. Soc.* **250**).

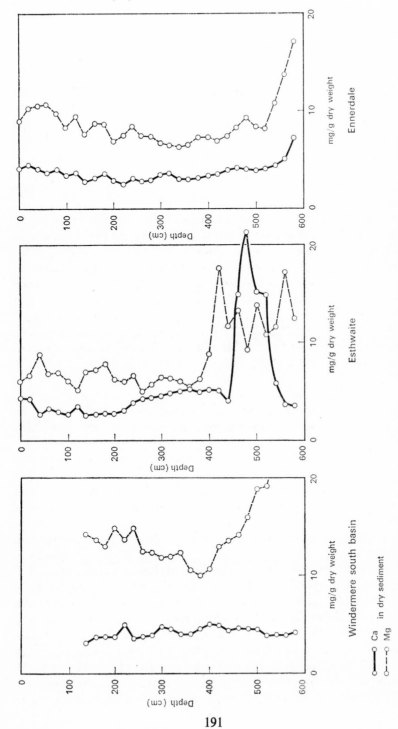

63. Distribution of concentration of calcium and magnesium with depth in the sediments of three lakes (Mackereth, F. J. H. (1966), *Phil. Trans. R. Soc.* **250**).

close relationship between the Linsley Pond profile and the general profile for the Lake District Valley lakes is coincidental seems hardly credible'. The explanation he believes, is to be sought in factors which operated widely over the northern part of the northern hemisphere in post-glacial time.

One interpretation of the carbon profile is that the initial rise corresponds to the deposition of vegetable remains on the bare soil left by the retreating ice, the plateau to a period of stability and the fall to a period of instability and erosion. Before this is accepted other possibilities are disposed of. A fluctuating rate of decomposition either on land or under water is unlikely because, as mentioned above, only the most stable compounds could have survived the vicissitudes to which they were subjected. It remains then to choose between the explanation offered and one based on variation in the rate of production of organic matter. If, argues Mackereth, the leaching-erosion one is correct, changes in the amounts of easily leached substances should clinch it. To this end he calculates the amount of sodium and potassium in the mineral fraction of the sediments. In a stable soil steady leaching should progressively reduce the proportion of these two substances, whereas under unstable conditions the removal of soil before it had been leached for long should cause an increase. This is what is actually found (fig. 62). The concentration, particularly of sodium, is high to begin with as the unstable soils left by the ice age are leached and eroded. Then there is a steady fall to about the critical date of 3000 B.C. followed by a rise.

Calcium (fig. 63) has a similar history, though its removal by leaching appears to have been greater than that of sodium. The high peak at the bottom of a core from Esthwaite is due to a concentration sufficiently high to cause precipitation in the water.

Mackereth had hoped that the concentrations of the halides and boron might reveal something about changes in climate, an increase indicating greater frequency and strength of winds blowing from the sea, as indicated in chapter 5. The expected increase at the time of the elm decline does appear in the cores, but it is not certain that it is not due to greater absorption of halides in the soil as leaching reduced the amount of bicarbonate derived from the rock.

Iron and manganese are of particular significance on account of the light they shed on events within the lakes (fig. 64). Both are relatively insoluble in the oxidized form and soluble when reduced, but manganese reduces at a higher concentration of oxygen than iron. In Ennerdale the ratio of the two has remained unchanged, and evidently the supply has depended on erosion. In Windermere, in contrast, there is a big rise in the concentration of manganese early in the post-glacial period, no doubt because the redox condition in the soils of the drainage area reached a point where manganese was reduced but iron was not. The manganese was

2. Esthwaite Water. The near horizon is the watershed.
The mountains beyond bound the drainage area of Windermere

3. Loweswater, showing a stony substratum in the foreground and a
reed-bed, mainly *Equisetum*, beyond

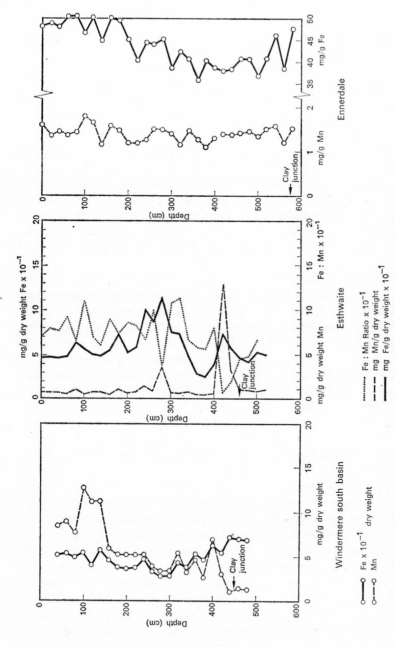

64. Distribution in depth of the concentration of iron and magnanese in the dry sediment of Windermere south basin, Esthwaite and Ennerdale (Mackereth, F. J. H. (1966), *Phil. Trans. R. Soc.* **250**).

precipitated in the lake. In the soils of the Esthwaite drainage basin the process went further still. There is an immediate increase in the amount of manganese at the start of the post-glacial period, but the high peak is followed by a fall to a level about half the initial one. It is postulated that this represents the onset of reducing conditions in the lake, which allow the manganese to dissolve in the water. It must be postulated further that it remains in solution until it is carried out of the lake and is not precipitated. A smaller peak at 280 cm below the sediment surface represents an increase in oxygen concentration sufficient to precipitate manganese, and thereafter there is a return to reducing conditions. Iron reaches its lowest level somewhat later than manganese, presumably because oxygen had by that time reached a level at which iron also was being reduced, dissolved and carried out of the lake. The concentration in the sediment then starts to rise, which is taken to indicate a rise in oxygen insufficient to oxidize manganese but sufficient to oxidize and precipitate iron in increasing quantities. At the culmination of this increase manganese is coming down as well but it is followed by a drop to levels at which manganese is permanently reduced but the degree of oxidation of iron fluctuates. The three peaks then represent three oxygen maxima.

Mackereth maintains that only by the postulating of oxidation and loss in the lake can the increase in the amount of manganese, relative to the amount of iron when the latter is at a maximum, be explained.

Some of Mackereth's conclusions about the lakes themselves are unexpected. The old idea that the history of a lake is one of enrichment as the rocks of the drainage area weather is shown to be the reverse of the truth, as Round (1957a) had already suggested. The lake waters were richest in mineral substances just after the ice had retreated, leaving the land covered with unstable soil. After this had been covered with vegetation it was subjected to a long period of leaching during which the supply of ions to the water, higher at first than at any subsequent period, fell gradually. The calcium content in particular was high in comparison with that of later years. It will be recollected that production of phytoplankton is not greatly influenced by the concentration of most of the common ions, and appears to depend more on the presence of some unidentified, probably organic, substance or substances. Whether this was present 8000 years ago is not known, but there seems no reason to doubt that soil then was similar to soil today and produced the same stimulants to the growth of algae. If this be so, the lakes were probably barren when they were first formed in spite of the rich supply of inorganic nutrients. However, as soil formed in the drainage area, and one must suppose that a warmer climate played a part too, production may well have reached a point from which there has been a decline, steady at first but irregular after man's removal of trees set off erosion. Only in the most recent times has there been an upward trend, due to enrichment from agricultural fertilizers and sewage.

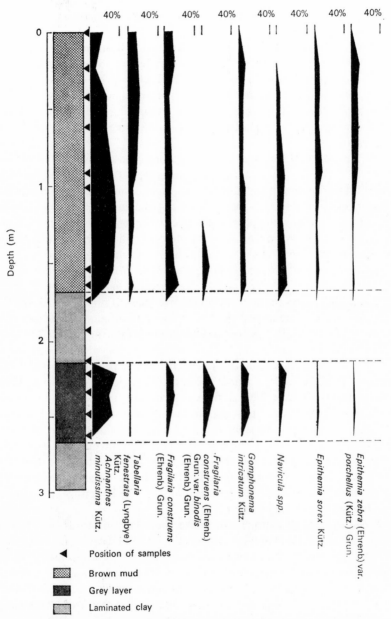

65. The percentage frequency of the seven most abundant species of diatoms through-out a core from very shallow water (3 m) (Pennington, W. (1943), *New Phytol.* **42**).

Esthwaite, if Mackereth's deductions are correct, has been eutrophic since early in the post-glacial, and the greatest oxygen deficiency occurred during the first half of that period. This had not been suspected by earlier workers.

If this reconstruction of past events from cores be likened to the putting together of a jig-saw puzzle, Mackereth may be said to be confronted with one whose pieces, though some have been chewed by the dog till their exact relation to the rest can no longer be made out, are at least all there. The biologists, on the other hand, are assembling a puzzle with many pieces missing. It is no matter for surprise that their final picture is less informative.

Pennington (1943) divides the diatoms into those which are small and numerous and those which are large and scarce. The small diatoms comprise littoral species in shallow water and planktonic species in deep. The percentage of the littoral species does not change much throughout the core (fig. 65), but of the planktonic species (fig. 66), there are two, *Asterionella* and *Synedra radians* that are almost confined to the surface and two, *Melosira italica* and *Cyclotella comta*, which are commoner in the clay bands than in the brown mud. *Melosira italica* has been regarded as characteristic of lakes rich in dissolved organic matter and *Asterionella* is found today only in the more productive lakes. If it is assumed that the rate of mineral sedimentation has been constant during the last few centuries, *Asterionella* has been abundant in the lake for the last 220 years at most. This gives a date of 1720 for its appearance, which is well before there was any invasion of the Lake District by people attracted to its beauty. It is, however, towards the end of a period of some sixty years which seems to have been a time of prosperity; the scarcity of older dwellings suggests that most people pulled down their houses and built better ones during it. Possibly improved methods of cultivation led to the prosperity and to the change in the lake that favoured *Asterionella*; there was an improvement in this century, though it started later than 1720 (Rollinson 1966). If rise of *Asterionella* was due to an increase in the human population and the enrichment of the lake by sewage, the method of dating must be inaccurate.

The large scarce diatoms are unlike the small numerous ones in two ways: there is little difference between shallow and deep water, and nearly all the species vary in abundance in different parts of the cores. Most are less abundant in the brown mud than in lower layers, but the reverse is true of a few, notably *Epithemia turgida* var. *granulata* and *Gomphonema geminatum* (figs. 67 and 68). All are living today, except *Melosira arenaria* var. *hungarica*, but none is now abundant.

Pennington (1943, p. 21) writes 'practically nothing is known of the ecological requirements of diatom species' and is unable, therefore, to derive conclusions from the findings.

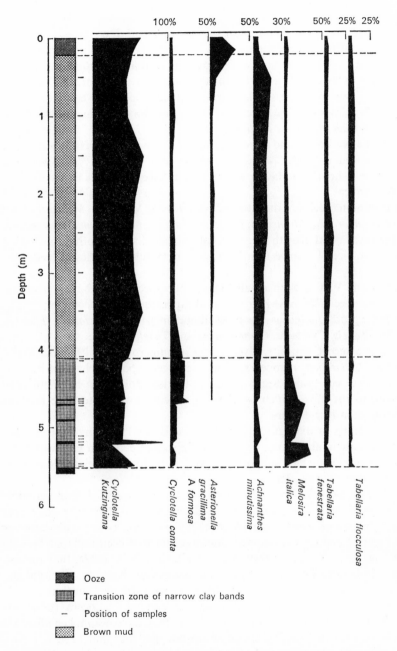

66. The percentage frequency of the seven most abundant species of diatoms throughout a core from deep water (65 m) (Pennington, W. (1943), *New Phytol.* **42**).

197

Round studied cores from Esthwaite (1961c) and from Kentmere (1957a), a basin that has filled up and is now a commercial source of diatomaceous earth. The lists of species from both these lakes fill several pages, and the two are compared with each other and with Windermere. The three are by no means identical but certain important generalities emerge. In zones I, II and III growth of diatoms was poor, and *Melosira arenaria* was the main species. Round quotes a personal communication from Pennington to the effect that this is probably indistinguishable from her var. *hungarica* (now *M. teres*).

In the pre-boreal and boreal, zones IV, V and VI, growth is still poor. The typical diatoms, *Melosira arenaria* and species of *Epithemia* and *Fragilaria*, flourish today in waters rich in bases, which is in accord with the chemical results.

Common species in the Atlantic period, zone VIIIa, are *Melosira italica* subsp. *subarctica*, *M. distans*, *Cyclotella comta*, and *Stephanodiscus astreae*. There was good growth of all except epipelic diatoms during this period and evidence of change from species of base-rich to species of base-poor waters.

In zone VIIb growth of all but epipelic species reached a maximum, but the species of base-rich waters disappeared almost completely. The following are quoted as the most abundant genera, *Cyclotella*, *Eunotia*, *Anomoeoneis*, *Cymbella*, *Gomphonema* and *Tabellaria*.

In the recent deposits, *Asterionella* is numerous and *Melosira italica* subsp. *subarctica* reappears. Some epipelic species characteristic of productive conditions are common in the lake today but have not been found in the mud, which is taken to indicate a sharp advance towards entrophy in recent years.

According to the figures given by Round (1961) and by Tutin (1943), the increase in abundance of *Asterionella* started earlier in Esthwaite than in Windermere, which supports the contention that it is related to cultivation rather than sewage.

Pennington and Frost (1961) record the discovery of four vertebrae and some scales from a fish in late-glacial deposits. However, they were in a core where there was evidence that the sediments originally covering the glacial deposit had slipped sideways and had been replaced by mud of later date. It is, therefore, not certain, though probable, that the fish lived in late-glacial times. The remains are clearly those of a salmonid and are stated to be most probably 'either *Salmo trutta* or *Salvelinus*'.

Goulden (1964) found thirty-seven species of Cladocera (table 41) in the sediments of Esthwaite, fifteen more than have been taken alive. He believes that the missing species have not died out and have not been recorded because there has been no intensive search for them. *Diaphanosoma brachyurum* is the only present-day species which has not been found in the sediments.

198

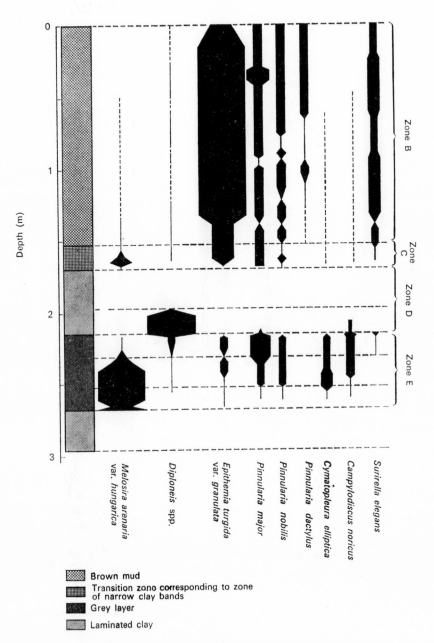

67. The vertical distribution (approximate frequency) of large conspicuous diatoms in a core from very shallow water (3 m) (Pennington, W. (1943), *New Phytol*, **42**).

Of the thirty-seven, twenty-three belong to the family Chydoridae and all these, together with *Sida* and *Polyphemus*, are littoral.

There have been five periods when littoral Cladocera were abundant and these are related to the climatic amelioration of the Allerød period, the similar change in the pre-boreal, the climatic change that marked the start of the Atlantic period, the Neolithic deforestation and the Norse

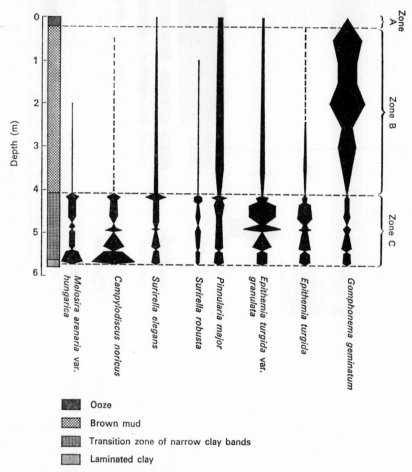

Ooze

Brown mud

Transition zone of narrow clay bands

Laminated clay

68. The vertical distribution (numbers per mg) of large conspicuous diatoms in a core from deep water (65 m) (Pennington, W. (1943), *New Phytol.* **42**).

deforestation. Whether the reduction in numbers in between represented a fall in production or whether animals that have not been preserved were abundant then cannot be determined.

There has been a change in the composition of the population with time which is attributed mainly to temperature. From a study of present distribution and occurrence in cores elsewhere, Goulden concludes that

Table 41

Species of Cladocera reported from Esthwaite Water sediments and from plankton collections (Scourfield and Harding, 1958). (Goulden, C. E. 1964. *Arch. Hydrobiol.* **60**).

	Sediment	Surface sediment only	Scourfield and Harding (1958)
Sida crystallina	X	X	X
Latona setifera	X	–	–
Daphnia hyalina var. *lacustris*	?	?	X
Daphnia (sp. 'A' ephippium)	X	X	–
Daphnia cf. *pulex*	X	–	–
Ceriodaphnia megalops	X	–	X
Ceriodaphnia pulchella	X	X	X
Ceriodaphnia quadrangula	X	X	X
Bosmina coregoni	X	X	X
Bosmina longirostris	X	X	X
Ilyocryptus sp.	X	–	–
Bythotrephes longimanus	X	X	–
Polyphemus pediculus	X	–	X
Leptodora kindtii	X	X	X
Eurycercus lamellatus	X	X	X
Camptocercus rectirostris	X	X	–
Acroperus harpae	X	X	–
Graptoleberis testudinaria	X	X	X
Alonopsis elongata	X	X	X
Alona affinis	X	X	X
Alona costata	X	X	–
Alona guttata	X	X	X
Alona intermedia	X	X	–
Alona quadrangularis	X	X	–
Alona rectangula	X	X	X
Alonella nana	X	X	X
Alonella excisa	X	X	X
Alonella exigua	X	X	X
Peracantha truncata	X	X	X
Pleuroxus trigonellus	X	–	–
Pleuroxus laevis	X	X	–
Pleuroxus uncinatus	X	–	–
Chydorus globosus	X	–	–
Chydorus piger	X	X	X
Chydorus sphaericus	X	X	X

Table 41—*cont.*

	Sediment	Surface sediment only	Scourfield and Harding (1958)
Anchistropus emarginatus	X	–	–
Rhynchotalona falcata	X	X	–
Total 'Non-Chydorid' species	14	9	10
Total 'Chydorid' species	23	19	12
Total species	37	28	22

Chydorus sphaericus, Alona affinis and *Acroperus harpae* (fig. 69), the three species that are commonest in the Allerød, pre-boreal and boreal periods and scarcer later, are very tolerant of low temperatures. *Alona quadrangu-*

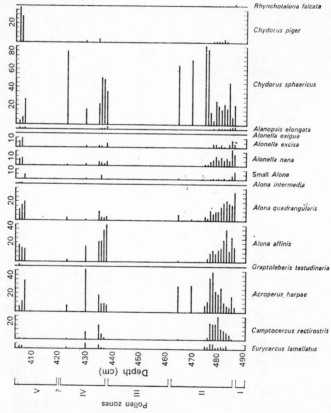

69. Percent composition of 'Chydorid' *Cladocera* in the Late Glacial and immediate Post-Glacial sediments of Core 2, Esthwaite Water (Goulden, C, E, (1964), *Arch. Hydrobiol.* **60**).

laris, *Camptocercus rectirostris* and *Alonella nana*, which have a smaller difference in abundance in early and late post-glacial times are less tolerant but can exist in cold regions. In the individual accounts of the species Goulden writes that, although *Acroperus harpae* is abundant only in the colder regions of the world, it is not confined to them and this 'implies that there are several distinct physiological varieties of this species'. *Chydorus sphaericus* is associated with blue-green algae, and its decrease in

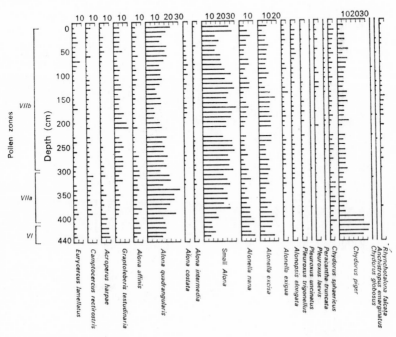

70. Percent composition of 'Chyorid' *Cladocera* in Post-Glacial sediments of Core 1, Esthwaite Water (Goulden, C. E. (1964), *Arch. Hydrobiol.* **60**).

boreal, Atlantic and post-Atlantic times 'suggests an absence of blue-green algal blooms during this time'. These remarks are quoted here in order to stress the difficulty, familiar to all zoogeographers, of reconstructing conditions in the past from remains preserved until today.

In the latter half of the post-glacial period a number of species not present earlier entered the lake (fig. 70).

Planktonic species have contributed more than 50 per cent to the remains almost continuously since the pre-boreal period. There has been little of note in the composition of the population until a level about 100 cm below the surface, when there were outstanding changes which may be discussed together with those in the proportions of certain chironomid genera (fig. 71). They are associated with changes in the environment brought about by the Norse occupation. *Bosmina longirostris* appeared at this point and increased steadily, while *B. coregoni*, which had been a common species throughout the previous period, declined. Latterly the proportions of the two species has been reverting. *Ceriodaphnia pulchella* and *C. quadrangula* appeared at the same time as *B. longirostris* and increased in the same way but reached a peak and started to decline later. *Daphnia* species have recently risen to a high peak at the same time as *Ceriodaphnia*. *Chironomus* first appeared before *B. longirostris* but was not abundant till a level about 55 cm below the surface at which time both *B. longirostris* and *Ceriodaphnia* showed a first peak. It then declined before rising to a second peak contemporary with those of *Daphnia* and

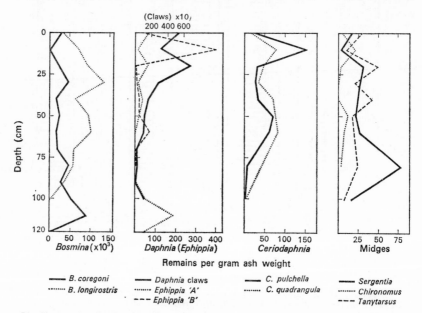

71. Changes in abundance of *Bosmina*, *Daphnia*, *Ceriodaphnia* and midge remains in the upper 120 cm of Core 1 sediment, Esthwaite Water (Goulden, C. E. (1964), *Arch. Hydrobiol.* 60).

Ceriodaphnia. The replacement of the one species of *Bosmina* by the other indicates the onset of more productive conditions and *Chironomus* belongs to a genus some of whose species are characteristic of lakes poor in oxygen. Goulden (p. 13) suggests that Esthwaite became anaerobic between A.D. 1000 and 1200. On page 35 he suggests that the decline of *Chironomus* and the three planktonic Cladocera mentioned indicates that oxygen depletion should no longer occur. He explains away Mortimer's record of de-oxygenation by postulating that it is now irregular and that Mortimer happened by chance to observe two unusual years in which it did take place. This must strike anyone familiar with the lake as inherently improbable, but there is also firm evidence, for which I am indebted to Dr V. G. Collins, that the hypolimnion of Esthwaite has become de-oxygenated every summer for the last twenty years. It will clearly be necessary to find out more about the animals before they can be used to challenge the conclusions of the chemists, and little weight can be attached to Goulden's contention that Mackereth is wrong when he states that Esthwaite was anaerobic in early post-glacial times because the animal remains do not provide any evidence.

CHAPTER 13

Bacteria and Fungi

When Dr C. B. Taylor started work at Wray Castle in 1938, knowledge about bacteria that might indicate contamination with sewage was extensive, but little work had been done on other types. The first problem was one of counting, the second a study of origin and distribution, and the third an investigation of activity. Taylor (1940) contributed to the first two of them and in later articles (1942a, 1949a) to the third. He also investigated the distribution of *Escherischia coli*, the indicator of pollution (1941a, 1942b). Four other papers (1939, 1941b, 1948, 1949b) are more in the nature of progress reports, being lectures delivered to interested societies or contributions to reviewing journals, though some contain original information.

Dr Vera Collins continued the work after Taylor's departure and also carried on the tradition of general articles (1957, 1963). In 1957 she described a method of counting bacteria directly in collaboration with Miss Charlotte Kipling, the Association's statistician. In 1961 she contributed to the second and third of the problems mentioned above, and in 1962 and 1966 collaborated with Dr L. G. Willoughby, who was studying fungi and the distribution of their spores. Their observations were made more frequently than previous ones and revealed such large fluctuations in such short spans of time that they supersede much of the earlier work on distribution. Accordingly these studies are described first, after the paragraph devoted to methods.

Taylor (1940) considered ways of counting bacteria directly but did not find a suitable one, and was obliged to fall back on culture, though aware that only a proportion of the bacteria present would grow on any medium into which they were inoculated. He tested seven media at various temperatures and found sodium caseinate agar at 20°C to be the most satisfactory; this was the medium used in later work. Collins and Kipling (1957) added glycerol and gentian violet to drops of water, and then placed them in a desiccator until only glycerol and stained bacteria remained. This liquid was then drawn into fine capillary tubes, which they had to make themselves, and the bacteria in each were counted under a microscope. The volume of liquid was ascertained by weighing. The number of bacteria revealed by this method ranged between six and eleven thousand times that found by culture methods.

Collins and Willoughby (1962) made their observations in Blelham Tarn, during the course of an experiment in which stratification was destroyed by means of compressed air which carried hypolimnion water up into the epilimnion through pipes located at various points. The sequence of events was as follows (fig. 72):

17 July. The thermocline lay between about 6 and 7·5 m and there was no oxygen below 8 m. This was the typical summer condition of this lake. Its greatest depth is 13·5 m.

31 July. Pumping started.

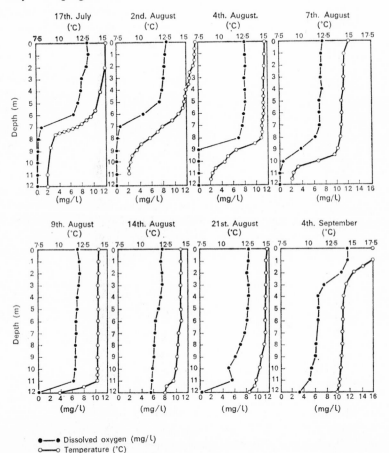

●——● Dissolved oxygen (mg/l)
○——○ Temperature (°C)

72. Dissolved oxygen and temperature profiles at Blelham Tarn (Collins, V. G. and Willoughby, L. G. (1962), *Arch. Mikrobiol.* **43**).

2, 4, 7 August. The thermocline had sunk a little deeper on each of these successive dates.

9 August. The hypolimnion had now been reduced to the bottom metre of that part of the lake when the samples were taken.

14 August. Oxygen was now almost uniform from top to bottom, though the bottom half-metre was a little colder than the water above it.

21 August. Both curves still showed traces of a hypolimnion at the very bottom of the lake.

31 August. Pumping ceased.

4 September. Oxygen and temperature indicated that the overturn had been complete and that stratification had begun to re-establish itself near the top.

The numbers of bacteria (fig. 73), determined by culturing, were related to the rainfall (fig. 74) as well as to the artificial overturn. On 17 July bacteria were abundant in and above the thermocline but scarce below it as well as in the topmost two metres of the lake. This is a pattern that had

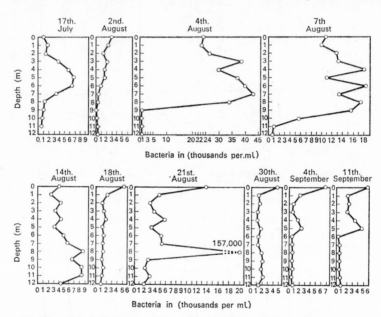

73. The vertical distribution of bacteria in Blelham Tarn (Collins, V. G. and Willoughby, L. G. (1962), *Arch. Mikrobiol.* **43**).

often been observed before, and the explanation offered is that the change in density prevents detritus and dead cells of phyto- and zooplankton falling through into the hypolimnion, and their accumulation in the middle of the water column provides bacteria both with the site for attachment which many freshwater forms require and also with a readily available source of oxidizable organic matter. On 2 August bacteria were scarcer. They showed a tendency towards regular decrease from top to bottom. This sample was taken just before an unusually heavy day's rain (fig. 74), the result of which was a colossal increase in the number of bacteria; it will be noticed that the scale for this month on fig. 73 is more contracted than for the rest. Also striking is the absence of any effect on the hypolimnion. This phenomenon of a big increase in numbers after rain had been

5. Gilson mud sampler

6. Jenkin surface mud sampler closed, from the side, with full tube in the foreground

7. Jenkin surface mud sampler open, with the tube to the front

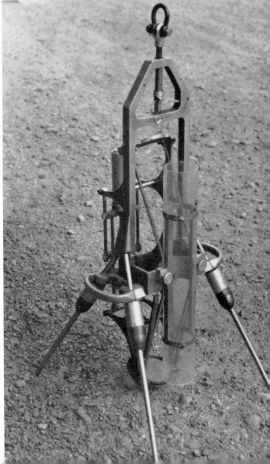

8. Vegetation sampler

9. The Jenkin corer.
(Mr Jenkin is nearest
the camera)

frequently noted before and is mentioned by Taylor (1940) and Collins (1961).

During the next ten days there was a fair amount of rain and a decrease in the number of bacteria together with a lowering of the depth to which they were plentiful as the thermocline sank. The sample on 18 August was

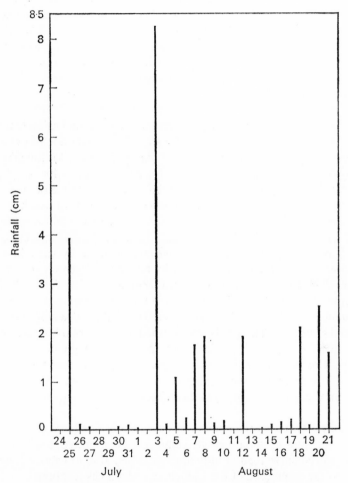

74. Rainfall (Collins, V. G. and Willoughby, L. G. (1962), *Arch. Mikrobiol.* **43**).

taken at the end of a period during which rainfall had been slight. It had been known that numbers of bacteria decreased during dry weather but the extent of the fall had not been demonstrated so dramatically before. It is impossible to state how much of the loss is due to the death of the bacteria and how many are carried through the lake. In this instance death must have been the greater source of loss because there was hardly any

P

stratification and, at the previous sampling, bacteria had been most abundant in the lower half of the lake.

There had been some rain by the time of the next sampling on 21 August and bacteria were extremely numerous in a narrow zone. The reason is not clear. Previous examples of this type of distribution had been attributed to stream water inserting itself between two layers of lake water of different density; generally it had come in warmer than the hypolimnion and colder than the epilimnion and had interposed itself between them. As the thermocline had been removed from Blelham artificially, this explanation cannot apply to it, but it is probable that the zone where bacteria were abundant is a zone of stream water confined between these levels for some similar reason.

Taylor (1949b) records that bacteria are at their most abundant in winter, the increase dating from the overturn and autumnal gales. Numbers fall in March and remain low until the end of stratification. The effect of rainfall on numbers is greater in winter than in summer. Willoughby and Collins (1966, figs. 33–36) confirm this as a general picture, but emphasize the great and rapid changes that can take place at any time of year.

A question of general importance concerns the extent to which the rise in numbers after rain is due to the import of bacteria from outside or to multiplication. If the latter takes place, there is a further and extremely difficult problem of how much multiplication is done by bacteria which have been washed in, and how much by bacteria indigenous to the lake. Taylor (1939) concluded that there must be multiplication in the lake from a consideration of the relative numbers in the inflow and in the lake itself. Willoughby and Collins (1966) compared the number of fungal spores and the number of bacteria. Spores were washed in by the rain but there could be no question of multiplication since spores do not give rise to spores directly. Comparison of spores and bacteria showed that the numbers are not always in the same proportion, bacteria being relatively more abundant when there are many of them. This is clear evidence of multiplication.

Taylor (1949a) used the amount of oxygen consumed in a given time at 20°C to measure bacterial activity (table 42). As expected, the figure was higher in productive than in unproductive lakes; also, as expected, having regard to the magnitude of the fluctuations which has just been described, the figures were not exactly in accord with the serial relations of the lakes shown by other studies. When glucose was added to the lake water, the increased activity was roughly proportional to the amount in untreated water. Addition of phosphate also led to greater consumption of oxygen, but no relation between the amount of the rise and any other factor could be detected. It was probably connected with the concentration of iron and other compounds which absorb phosphate, for this substance produced a greater effect in a synthetic medium than in any lake water. Taylor con-

Table 42

Bacterial activity measured in terms of oxygen consumed (p.p.m.). Comparison of lake water which has not been treated with that to which glucose and phosphate has been added (Taylor, C. B., 1949a. *Proc. Soc. appl. Bact.*)

	En	Bl	Lo	Es	Synthetic medium
Untreated lake water	0·16	0·51	0·69	1·05	–
Lake water + glucose	0·95	1·74	2·43	3·14	0·10
Increase caused by glucose	0·79	1·23	1·73	2·09	–
Lake water + glucose and phosphate	3·34	2·53	4·90	4·87	5·98
Increase caused by phosphate	2·39	0·79	2·48	1·73	5·88
Increase due to glucose and phosphate	3·18	2·02	4·21	3·82	–

En = Ennerdale, Bl = Blelham, Lo = Loweswater, Es = Esthwaite.

cludes that bacterial activity is generally limited by the availability of a source of energy but that, if more organic material were available in lakes, activity would be limited by the concentration of phosphate. The nature of the substances which, brought in by rain, make growth possible is still unknown. Taylor (1949a) records that the death of the cells after an outburst of a blue-green alga is followed by a big rise in the number of bacteria, but the death of a big population of *Asterionella* by only a small one. He points out too (1949a) that some bacteria are active only when attached to something, and the effect of heavy rain is to increase the number of supports suitable for such species.

Taylor (1942a), testing some 800 cultures of bacteria from natural waters found that most are rod-shaped and Gram-negative, that many are chromogenic, and that cocci and spore-reproducing forms are rare. In this they differ from samples of soil-bacteria. When subjected to the classic biochemical tests of bacteriology many gave no reaction and Taylor doubts whether these tests are likely to be useful in classifying lake bacteria. Dr Collins, however, informs me that this view is no longer tenable, in view of recent work which improved methods of culture have made possible. It has not been published but is referred to in the thirty-third annual report (1965) of the Freshwater Biological Association.

Some of the results presented by Collins (1961) have already been noted. In addition, this paper describes a study of the bacteria of lake muds, many of which are found also in surface waters during full circulation in winter, when they are evidently swept up in currents caused by the wind. Collins (1963) gives a valuable review of the whole field, and includes in an appendix an account of the culture-methods used. In the mud were found bacteria which: produce ammonia from peptone, nutrose and urea; produce sulphuretted hydrogen from peptone; liquefy gelatin; reduce nitrate and nitrogen; fix nitrogen; ferment dextrose; ferment xylose; hydrolyse starch; decompose chitin; decompose cellulose; hydrolyse fat; and reduce sulphur. There is only a little fixation of nitrogen and there is a considerable amount of hydrolysis of starch, but the tremendous task of obtaining exact quantitative data for all these processes at different times of year has not been undertaken.

Collins suggests that the following sequence of events takes place. After stratification aerobic bacteria reduce the concentration of oxygen in the mud. When it is low, sulphate-reducing bacteria become active, and the resulting hydrogen sulphide and ferrous sulphides, diffusing into the water, deplete the store of oxygen there. *Chromatium*, a coloured form which can use the sun's energy to oxidize sulphides to sulphate, is found at the boundary between water with and water without oxygen, and in Esthwaite it has been found as deep as 13 m. In the upper layers, where there is oxygen and organic matter, members of the family Athiorhodaceae are abundant, together with iron bacteria such as *Sphaerotilus natans, Leptothrix ochracea, L. crassa* and others. This stratification is destroyed at the overturn and many of the organisms are found at all depths. Thiosulphate-oxidizing bacteria in the mud and just above it complete the sulphur cycle.

Little was known about the aquatic fungi when Dr L. G. Willoughby came to the Windermere Laboratory to start work on them in 1956. Much of his work has been taxonomic and an account of it therefore lies outside the scope of this book. Spores are numerous in the lakes after heavy rain though they disappear soon in dry weather. Tracing their origin involves culturing on specially prepared media. Some originate from truly aquatic forms, many originate from sites at the edge of the lake which are uncovered when the water level is low, and some originate in the soil (Willoughby 1962, 1965). Spores that fall into the hypolimnion when oxygen concentration is low do not survive long. In Blelham, Willoughby and Collins (1966) found that *Aureobasidium pullulans* was a common species particularly numerous in Ford Wood Beck and Wray Beck. The point of origin of many of the spores was found to be oak and sycamore leaves. Over eighty species of Saprolegniaceae were isolated from Blelham and the main source of some of them was a small trickling filter purifying the sewage from the village of Outgate and discharging into Ford Wood Beck.

CHAPTER 14

Methods

It can be argued that in accordance with the policy governing the writing of this book, methods are of general application and should, therefore, be omitted. On the other hand, the method used is an essential part of any piece of research and one should not be included without the other. A list of references, at least, is therefore indispensable. Some readers may feel that they do not wish to read about a piece of apparatus unless they might want to use it, and that therefore the short descriptions without constructional details, all that there is room for here, are of no value. However, it is not always clear from the title whether a paper is concerned narrowly with one particular experiment or whether it reviews a wide field, as several of the papers about to be mentioned do. It has been decided, therefore, that an annotated list is preferable to a bare one, and that is the form which this chapter takes.

Of the Freshwater Biological Association's Scientific Publication No. 21 (Mackereth 1963) on chemical methods, the Director, Mr H. C. Gilson, writes in his foreword: 'Most of the more sophisticated methods described are those in regular use by Mr Mackereth and Mr Heron at the Ferry House, but we have also retained descriptions of the methods that can be used with simple apparatus.' Two papers not mentioned in this booklet are about a colorimetric method for routine estimation of calcium (Mackereth 1951) and about the estimation of calcium and magnesium in waters of low alkalinity (Heron and Mackereth 1955).

The importance of an instrument that will record oxygen concentration continuously has long been a challenge to the inventive. Hitherto one difficulty has been to keep the electrode free from the deposition which, inseparable from the passage of an electric current, generally gradually insulates the electrode. One solution to this problem is a continuously renewed surface, as in the dropping mercury electrode, but this involves cumbersome apparatus. Mackereth (1964) used an electrode of granulated lead and found that the lead hydroxide precipitated did not impede the flow of ions to it. When the other electrode is silver, a current flows in the presence of oxygen. Electrons travel from the lead to the silver electrode and there, in the presence of oxygen, form hydroxyl ions. These travel through the electrolyte to the lead electrode where the lead hydroxide is precipitated.

The electrodes are bathed in an aqueous solution of saturated potassium hydrogen carbonate, between which and the external medium is a polythene membrane that permits the passage of gas but little else. The rate at which oxygen passes through the membrane depends on the permeability of the membrane, which is known, and on the concentration outside. The amount of oxygen inside determines the amount of current flowing between the electrodes and this, therefore, can be used to discover the concentration in the external medium.

Mackereth (1963) includes a section on simple methods of collecting samples. Water may be obtained from any depth by lowering a tube of the required length and then sucking. The idea of turning a bicycle pump into a suction engine by reversing the washer was conceived by Dr C. H. Mortimer in the early days at Wray Castle. It may have proved a boon to subsequent generations but was viewed in a different light at the time by those of his colleagues who rode bicycles. It is recommended that the water be sucked through two bottles, of which the one nearer the pump is some three times the size of the other. When this larger bottle is full the smaller one should have been washed through sufficiently to provide an uncontaminated sample. Water can be brought up by means of an air-lift pump, which is merely a stream of bubbles released just inside the lower end of a larger tube from a smaller one. If the column which it is desired to sample is not too deep, a tube armed with enough weight to keep it straight and a string with which to pull it in provides a simple method of sampling. It is lowered to the required depth, the upper end is then closed and the lower end is hauled up by means of the string. This simple device, described by Lund (1949a), is particularly useful for samples of phyto-plankton. Samples from deep water are generally obtained with a closing water bottle.

The water bottle devised by Mortimer (1940) for bacteriological samples is more logically described at this point rather than later in conformity with the order in which the subjects were taken in the preceding chapters. Enquiring into current methods, Mortimer found two main categories, one involving an evacuated container whose neck could be broken off at the desired depth, the other a bottle with two tubes whose outer ends could be opened to obtain a sample. Water enters through a long tube which runs nearly to the bottom of the bottle and air is driven out through a short one that does not project below the inner end of the cork. The preparation of an evacuated container is time-consuming. Mortimer, testing the second method, found that when the tubes are straight, as those of some previous workers had been, a small amount of water enters as the bottle is hauled upwards. This could be prevented by bending the outer part of the short tube into the shape of an 'S'. Mortimer used copper tubes and closed them by a link of glass rod. This was attached to a spring kept stretched by a standard releasing-hook at the end of a cable, the whole being in a metal frame. When the bottle was at the required depth, a messenger sent down the cable released the hook, which left the spring free to contract and pull out the stopper.

Mortimer, as already noted, attached a string of thermistors slung from a buoy to a recorder in the laboratory and thereby obtained a continuous record of the temperature at nine different depths in Windermere. The use of thermistors is discussed by Mortimer and Moore (1953), and the subject of measuring temperature in water is reviewed by Mortimer (1953). Measurement of light is the subject of a long section in a review of methods with special reference to algae by Lund and Talling (1957).

The Utermohl method, which consists of counting, by means of an inverted microscope, algae which have been sedimented in a tube with a very thin bottom, has been used extensively in phytoplankton work. Lund, Kipling and Le Cren (1958) describe the method in detail and give an account of statistical procedures connected with it. Lund (1959a) describes a simple counting chamber for nannoplankton. Two strips cut from a long thin coverslip are glued to a microscope slide parallel with the long edges. A similar coverslip is glued on to them. The depth of the chamber so formed can be discovered in four ways. It can be measured directly by means of the fine adjustment of the microscope, if this is accurate, which often it is not. It can be discovered from area and volume, the former being measured on squared paper and the latter determined from the weight of the chamber when it is dry and when it is full of distilled water. The third method is to make direct measurement with an engineer's dial gauge attached to the microscope. Fourthly it can be calculated from counts of the number per field of particles whose number per unit volume is known. Reverting to the topic after six years' experience, Lund (1962) recommends the second method. Area should be determined not only by direct measurement but also planimetrically. Polystyrene has been found to be the most satisfactory cement.

Although little work has been done on zooplankton, Ullyott (1937) did devise for counting it an apparatus that has been found useful in the quantitative examination of other types of sample in which the animals are small. It consists of a trough about 30 cm in length and in width a little less than the field of the low-power binocular microscope to be used. The bottom of the trough is of transparent material. The trough fits into a frame or on rails in such a way that the entire length of it can be pulled across the field of the microscope. The traction must be of such design that the trough can be started and stopped without a jerk. Originally the source of energy was a weight, and the trough was set in motion by pressure on a pedal which took the brake off a big flywheel. On a later version depression of the pedal pulled the trough along and, when it was released, a weight pulled the fly-wheel back to the resting position. As it was connected to the driving shaft by a pawl and ratchet, it moved the trough only when pulled down. Subsequently traction was provided first by a clockwork gramophone motor and then by an electric motor, though whether these modern and more complicated machines brought increase in efficiency is not certain. The brake in later models was operated by sideways pressure of the knee. The leg was always used in order that both hands might be free, one to focus the microscope and the other to tap the keys of a counting machine.

Moon (1935a) used what he calls a scoop for collecting in water down to a depth determined by the length of his thigh boots. In fact it differed from a net mainly in its strong frame and cutting edge. For deeper water he arranged stones on a tray, reproducing as best he could the conditions of the lake substratum. The tray was then lowered down to the bottom of the lake, left for a period which trials had shown to be sufficient for colonization, and then removed for examination. His quantitative collections of *Asellus* made some years later were based on collecting for a known time. This method has been much in vogue recently, and was used by, among others, Mann and Reynoldson, in their extensive studies of leeches and flatworms respectively. Macan (1958) has reviewed methods of sampling in running water, and some of the techniques can be used on stony substrata in lakes too. Another review in the same series is that of Mundie (1956) on methods of trapping aquatic insects as they emerge.

When a bed of vegetation is sampled by means of a grab, the plants are pushed downwards and severed only when pressed between the substrate and the edges of the sampler. The volume of the contents of such a sampler may not be the product of the depth of the vegetation and the area of the grab. In an attempt to ensure that they are, Macan (1964) devised an instrument, which, as it is lowered into the vegetation, cuts each leaf or stem as it comes to it. Two tubes, each provided with teeth at the lower end, fit one within the other with the teeth at the same level (Plate 8). The outer can be revolved around the inner over a short arc, an operation which causes the teeth to pass to and fro across each other and sever anything which comes between them. A stud on the inner tube passing through a slot in the outer keeps the two together and makes the limited rotation possible. The sampler is pushed into the substratum to make a seal and, before it is withdrawn, the top of the inner tube is closed with a cap consisting of two hard discs with a rubber one between them. Tightening of the nut on a screw which passes through all three compresses the rubber and makes a seal. The diameter of the inner tube was 10 cm, the length of the outer tube 1 m and that of the inner tube 50 cm. This sampler has proved satisfactory in a pond but is of limited application in a lake because soft substrata in shallow water are rare.

Mud is easily sampled with a Birge-Ekman grab. A small sample can be collected most simply with an open tube, especially if the top can be closed before the sampler is pulled out of the mud. Mr H. C. Gilson has designed a simple piece of apparatus which has not been described but which is supplied by the Freshwater Biological Association (Plate 5). It is very convenient for use in depths at which it can be forced into the mud on the end of a pole, and it is in fact fitted with a screw to which the segments of the poles provided with pond-nets can be attached; it can also be used in deeper water, where it is driven into the mud by means of weights. The top of the cylinder is perforated and below it is a plate which, when forced against the top by means of a spring, seals the cylinder; the whole is somewhat reminiscent of the plug and plug-hole of a bath. The plate is on the end of a tube which fits inside another attached to the top of the cylinder. A compressed spring tends to push the inner of these two tubes upwards and thereby close the top of the sampler. It is kept open by means of a pin passing through both tubes. The pin has a head like that of a nail. The main hole through both tubes is large enough to allow this head to pass. When the head projects beyond the outer tube release of the manual pressure necessary to keep the holes aligned forces the pin into a notch in the outer tube. This notch is of such diameter that it receives the shaft but prevents the passage of the head. The pin thus holds the inner tube against the spring and keeps the sampler open. When pressure is exerted, the inner tube is driven downwards, and the pin moves until its head is clear of the notch and opposite the hole, through which it is immediately drawn by a spring on the opposite side. When pressure is relaxed, the inner tube is now free to pull the plate against the top of the sampler and close it.

A sampler of the type just described, which does not allow completely free passage of water through it, has to some extent the nature of a solid cylinder when approaching the bottom. If the sediment is fine and flocculent, some of it may be pushed to one side. Studies of the surface between mud and water require a sample disturbed as little as possible, and to obtain such a sample Mr B. M. Jenkin designed what has come to be known as the 'Jenkin Surface Mud Sampler' (Plate 6). Mortimer (1942, p. 148) outlines its working principles and promises a full description elsewhere. This has not been published but construction plans are available from the Freshwater Biological Association. The sample is taken in a glass or perspex tube about half a metre long and 7·5 cm in diameter. The tube is borne on one side of a four-legged frame so designed that it sinks into the bottom until the tube is about half full of mud. It enters the mud open above and below. Lids to close the two ends are held on arms which are clipped back on the opposite side of the frame against the pull of strong springs. When the sampler has come to rest, a messenger releases a conventional trip mechanism which sets free the arms. These, pivoted at the centre of the frame, are pulled through an arc by the springs and brought to rest when the plates which they carry are pressed against the ends of the sampling tubes. It is necessary to have strong springs to keep the caps firmly in place, but it is desirable that they should be moved into position slowly. This is achieved through the medium of a piston attached to one arm on the side distant from the sampler tube inside a cylinder attached to the other. As the arms rotate the piston moves through the cylinder pushing

217

water out on one side and drawing it in on the other. As the hole for the egress of water is small, the piston and cylinder and the arms attached to them can move only slowly.

Work on fish has provided less scope for the inventive and ingenious because methods of capture have been of commercial importance for so long: also because there are many workers in the field and marine and freshwater problems are similar. However, the coming of the electric shocker opened up a new chapter in the field of capture, and Moore (1954) has made a contribution to it with an account of a portable device capable of catching small fish. Le Cren (1954) describes a subcutaneous tag for marking fish and presents data on which the merits of various systems can be compared. Moore and Mortimer (1954) describe an electrical instrument for detecting subcutaneous tags.

In contrast the design of apparatus to obtain cores has provided wide scope for inventive talent, for none of the instruments already known were found to be suitable (Jenkin and Mortimer 1938). First in the field was Mr B. M. Jenkin, already mentioned in connexion with the Surface Mud Sampler. Mr Jenkin was a retired engineer to whom the Association was also indebted, when work first started, for the ingenious and economical adaptation of the engine and transmission of an old car to drive a launch and also to haul up its anchor and a wire to which apparatus could be attached. The essential part of his coring apparatus (Jenkin, Mortimer and Pennington 1941) was a metal tube, one-third of whose circumference had been removed and replaced with a flat face. Inside this tube was another of smaller diameter and also with part removed. It was pivoted in such a way that it could be revolved outwards through a slit in the larger tube until its leading edge came up against the face-plate on the side away from the slot. The ends were closed with plates, and therefore, when it was everted after the larger tube had been driven into the substratum, it enclosed a half cylinder of mud. The apparatus was lowered from a pontoon moored with four anchors (Plate 9). When necessary, it was driven into the mud by the process of hauling up a weight attached to the same cable and letting it fall. When the sampler was at the desired level in the mud, a messenger was sent down the wire to release a bridle which, when tension was applied, rotated the half cylinder which enclosed the sample. A series of samples at different depths could be obtained by adding lengths of tube above the corer.

This was an ideal way of cutting out a core of mud, as there could be no compression and any smearing of the surface was in the plane in which the sediments had been deposited. The disadvantage was the weight of the apparatus and the fact that it required a pontoon. This restricted the number of places where it could be used.

Accordingly Mackereth (1958) turned his attention to the design of a lighter more portable apparatus (fig. 75). The principle is to hold an outer tube (A) above the mud surface and then, by admitting compressed air into the top of it, to force a tube inside it (B) down into the mud. The firm base which this requires is provided by a metal cylinder (G), which

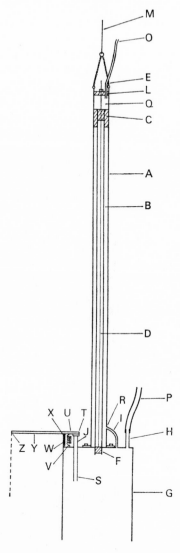

75. A cross-section of the Mackereth Corer (not to scale) (Mackereth, F. J. H. (1958) *Limnol, Oceanogr.* **3**).

has the form and dimensions of a domestic dustbin, though Mackereth prefers that it is referred to as the anchor chamber.

The apparatus is lowered on the end of a cable till the anchor chamber comes to rest on the bottom. Water is then pumped out of it through pipe P, an operation which causes it to sink into the mud where it provides a firm stable base for the rest of the process. Compressed air is now allowed to expand chamber Q between the top of the outer tube and the piston C which caps the inner tube. A third, fine, tube D passes through and is secured to the lid of the outer tube (L). Its lower end is attached to a piston (F) which fits the inner tube. As the inner tube is pressed downward the air inside it escapes through tube D.

When the inner tube has nearly passed out of the outer tube, a process whose completion is prevented by a flange, the compressed air passes through a side pipe (I) into the anchor chamber. This is lifted out of the mud and, once it is clear, its buoyancy is reduced through the valve S so that it does not reach the surface at a dangerous speed. Water forced into tube D pushes the inner tube back inside the outer and the piston F pushes out the core, which is 6 m long.

CHAPTER 15

Conclusion

Pearsall envisaged lake evolution as a process of increasing productivity due to the weathering of the rocks in the drainage area. This view has been discarded as research on cores, notably the chemical analyses of Mackereth, has revealed facts unknown to Pearsall when he first formulated his ideas.

The concept of a series, however, has been fruitful. Pearsall himself showed that the species of algae, rooted plants and fish change in a regular way along the series, and more recent data have not made necessary any alterations of the arrangement. Lund's calculation of production by algae is the most cogent confirmation of it and may remain so until figures for primary productivity are obtained. Little was known about the invertebrates in the early days but recent work has shown that they too fit well into the scheme.

The continental approach until recently relegated lakes to categories, and under its influence Mortimer stated that Esthwaite, in whose hypolimnion oxygen disappears in the summer, is fundamentally different from lakes in which this does not take place. For the student of organisms living below the thermocline it may be, but for the botanist or zoologist investigating the more diverse communities in the epilimnion, it is not. It is true that Pearsall found in Esthwaite certain communities of rooted plants that he found in no other lake, that Lund found there a greater predominance of blue-green algae than elsewhere, and that this lake is reputed to contain a greater abundance of coarse fish than any other. Whether all this amounts to a fundamental difference is a matter for dispute. The serious objection is that there are other eutrophic lakes in the Lake District in which these differences are not found. Lund showed that the production of algae in the South Basin of Windermere is second to that in Esthwaite and higher than that in other lakes in which oxygen disappears in summer. These also come lower in the series in an arrangement based on the other criteria which have been discussed. The scarcity of certain Ephemeroptera and Plecoptera on the stony substratum of Esthwaite affords a strong contrast with every other lake, but parts of Windermere harbour the same community. It is concluded, therefore, that the epilimnion is not greatly influenced by events in the hypolimnion and any scheme of classification based on them is not applicable to a whole lake. A better understanding of the communities in the epilimnion is to be gained from an arrangement of the lakes in a series than from a system in which they are separated into categories.

221

When discussing the occurrence or the abundance of individual species, Pearsall frequently recognized groups of lakes in preference to assigning a distinct place in the series to each one. The same treatment has been found convenient for the invertebrates. One outcome of this approach is that the dividing lines come in different places when different groups are considered. This may be illustrated by two invertebrate groups that have been studied thoroughly. There are a few more species of mollusc in Windermere than in other lakes but apart from this the same seven are found from Esthwaite down to Derwentwater, after which the number falls abruptly to four in Crummock, three in Buttermere and two in Ennerdale and Wastwater. The corixid fauna, in contrast, is similar in Esthwaite and Windermere and from Loweswater to Ennerdale with Ullswater and Coniston falling into an intermediate group. A study of tables in chapters 6, 8, 9, 10 and 11 would provide further illustration of this point. There are hardly any two lakes in the series between which dividing lines could not be drawn if the appropriate animal or plant group were selected. This is a further argument in favour of the serial arrangement in preference to one in which rigid categories are set up.

It has long been customary among biologists to regard lakes as units, though hints that they might not be have been coming from various quarters for some time past. Valovirta's demonstration that production in the bottom sediments of a Finnish lake decreases with increasing distance from the main inflow has been mentioned. Rodhe (1958), discussing the precautions that must be taken to ensure that C^{14} gives a reliable measure of production, points out that a change of wind will bring a different water mass to the sampling point and in it a different assemblage of phytoplankton. In the Lake District it was soon recognized that the two basins of Windermere differ sufficiently to be treated as two lakes. The work on the invertebrate fauna of the stony substratum of Windermere (fig. 54) underlines this point; there is a steady change from a biocoenosis in which certain insects are abundant, and flatworms and *Asellus* scarce or absent, to one where the proportions are reversed. The change is complete within one basin and, under these circumstances, the lake provides the ecologist with no more than a convenient boundary, and the unit which he must study is the biocoenosis.

The temptation to look into the future is always strong. The reward in terms of reputation is great for the occasional prophet who has the luck to guess right. There are no penalties, for wrong guesses are too frequent to attract attention. A correct forecast about the future of the Lake District lakes is more unlikely than many because the factors are political rather than biological. If the present trend continues, it is probable that, in current political jargon, the rich will get richer and the poor will get poorer. Marked improvement in the agriculture in the narrow valleys of the unproductive lakes is not likely, and further colonization by man will be

222

prevented by the Lake District Planning Board. One may suppose then that the process of leaching of the drainage area and the consequent impoverishment will continue. The productive lakes will become more productive as more sewage flows into them. Owing to the restrictions upon new houses imposed by the Planning Board, increase in the permanent population is likely to be slow, but facilities for the day visitor and for campers are likely to be improved. Existing houses at present served by earth closets or septic tanks will be connected to the main sewer. This may ultimately lead to the growth of unsightly algae which, piling up on the shore and decaying with an offensive odour, cause a public outcry that something be done. Advances in sewage purification may be made before this stage is reached. Investigation into methods of removing nutrients from sewage effluents by running them through tanks in which algae are grown are in progress at present, and more fundamentally new ideas have been put forward. It is possible, then, that the productive lakes will become less productive in the future. Whatever happens the ecologist will be in a position to study an environment that is changing fast and to take advantage of a situation which is in effect an experiment on a gigantic scale.

The problems that follow the enrichment of a lake, though familiar in several other countries, are unknown to most of those who strive to preserve the beauty of the Lake District and its amenities, and no emotions are aroused. The main target for hostility are those who wish to take water from the lakes for industrial and domestic purposes. Towards the end of the last century Manchester Corporation acquired Thirlmere and much of its drainage area, raised the level by means of a dam, and ordered the general public to remain outside a wall which they erected. It was a procedure that was commonplace among landlords at that time. Heavy taxation has greatly reduced the number of large landlords since then, and many of those that have survived have accepted changing public opinion about the privacy of wild places. A large corporation is not affected by taxation and, impersonal and remote, is less influenced by public opinion. The result has been strong feelings of antagonism towards Manchester Corporation. Lovers of beauty complain that the dam and wide zone of bare stones fringing the lake when the level is low detract from the magnificence of the scenery, and there has been violent controversy about the conifers planted to prevent erosion of the mountain slopes above the lake. However, it is probably the persistence of the 'keep out' notices that really annoys most people. Manchester acquired Haweswater before the war but, when they sought permission by means of a parliamentary Bill after the war, when the Lake District had become a National Park, to take water from Ullswater and Windermere, it was refused. The Bill went first to the House of Lords where the eloquence of Lord Birkett, the leading advocate of the day, caused it to be rejected without, according to the corporation, a hearing being given to the arguments in its favour. The corporation then made a different approach, which involved a public hearing at which eminent lawyers were retained by both sides. It was made clear at once that fears that access to the lakes would be restricted or that the levels would be altered were groundless; all that Manchester Corporation wished to do was to pump flood water out of the two lakes before it ran to waste in the sea. Pumping would cease when the lake fell to a level well above that reached in times of drought. There remained a further cause for mistrust. It was felt that, once the pumps were installed, further abstraction could be effected cheaply, and, in due course, as the demand for water grew, Manchester would return to ask permission to take more. This rational apprehension was soon shown to be groundless, for it was pointed out that there happened to be an amount of spare capacity in the aqueducts to Manchester that would just be filled by the amount that it was desired to take from the two lakes. More water could not be taken cheaply, as was feared, because more water would require another aqueduct all the way to Manchester, which would be extremely costly. These arguments carried the day, and the Minister concerned signed an order giving Manchester the permission requested. However, to make certain that the terms were adhered to, he stipulated that the water from Ullswater should run over a sill at the level at which pumping from the lake should cease, and that it should be pumped from the outflow side of the sill.

When this extraction is in operation, the only change will be that after much rain the lake level will subside a little more rapidly than it does now. It is difficult to see how this can affect the lake in any way. The Minister's stipulation seems to indicate a grudging assent, which should not encourage further plans for abstraction. Nonetheless plans, some of them much more undesirable because they involve an alteration in level, are being put forward. Their chances of success are impossible to estimate. There were those who hoped that, when National Parks were established, they would be secure from exploitation of all kinds. This was probably always optimistic in a country so thickly populated as Britain, but the impression gained is that all governments, when faced with the admittedly very difficult problem of setting a value upon something as intangible and elusive as recreational need and balancing it against physical demands upon which a price can be fixed, have not valued the first as highly as might reasonably have been hoped. Their future decisions are likely to depend on such imponderables as the intensity of demand, the financial state of the country, the availability of other sources of supply and perhaps also the personal attitude of a minister or the importance of other problems that a government is trying to solve at the moment when a decision has to be taken. Whether the Lake District lakes are left in their present state hangs on such factors as the financial feasibility of removing salt from sea water and the availability of money for projects such as the creation of a vast reservoir by damming Morecambe Bay.

Another temptation which is strong is to pontificate about the course that research should take in the future. There are many arguments in favour of resisting this temptation. The wisest counsel may be rendered void by advances in research and, therefore, what may be legitimate to lay before a committee at a given moment should not be committed to the permanence of print. One of the delights of a research career is that one is free to pursue one's own ideas and to take no notice of those of other people, particularly such as have the arrogance to attempt to dictate what should be done.

It is the author's hope, however, that a record of impressions gained from a study of so much of the work that has been devoted to the Lake District lakes will not be wholly unacceptable. The main impression is that a number of lines have been pursued with diligence and often with flair, but in isolation. Probably inevitable in the early stages, this isolation must be reduced sooner or later if the economy of a lake is to be understood completely, and research must extend sideways, till the various lines join, as well as progressing forwards.

No part of the aquatic ecosystem has received more attention than the phytoplankton, and no part less than the zooplankton. It is inevitable, therefore, that at present only the most tenuous ideas about the relation between them exist. It is a welcome sign of a less restricted attitude that a study of predation on phytoplankton has been started by a botanist, Dr Canter. The traditional gulf between botany and zoology originates in their treatment as separate subjects, an approach to biology that has been abandoned by several universities founded recently. At present it is accepted that a zoologist studies the food of a herbivorous animal and that an ichthyologist studies the food of a fish. As the idea gains ground that there is no gulf between the two subjects, and that a worker familiar with algae is the best person to study what is eaten by a herbivore, considerable precision in the results seems to be more than a possibility.

To turn to another food chain, the larger invertebrates have been studied but they too have been left in a limbo in terms of prey and predators. The latter have been the domain of the ichthyologists and, in the lakes at least, there has not been a fine correlation about what fish are eating and what is available in different parts of a lake for them to eat. The gap in knowledge about the other end of the chain is much greater. Information about bacteria is extensive. They may be an important source of food for many animals but the only work on this was done by Dr Webb during visits from Leicester University. Her studies of the Protozoa in Esthwaite linked them and the bacteria, but nobody has added any more to the chain. The Protozoa too are likely to be an important food supply, probably to some animals at all times, almost certainly to others when they first emerge from the egg and are very small. A great change in the biocoenosis on the stony substratum of Windermere was attributed to changes in the food supply. Bacteria and Protozoa are probably involved, but at present these remarks are speculations and the work that will provide a basis of fact remains to be done.

Biologists have chosen to work apart from each other, but have collaborated eagerly with chemists and physicists, which is probably an indication of the channels into which their thoughts were directed by those who taught them. The detached scientific observer may ask at this point whether the intellectual eminence or the social charm of the chemists and physicists compared with that of the biological members of the staff could have been a factor too. At this point the author, who hopes to continue to enjoy the same happy relations with all his colleagues as in the past, finds that his train of thought has led to a point where the prudent course is to write:

THE END

REFERENCES
1. Works concerned with the Lake District
(for other references cited see p. 239)

ALLEN, K. R. (1935). 'The food and migration of the perch (*Perca fluviatilis*) in Windermere.' *J. Anim. Ecol.* **4**, 264–73.

ALLEN, K. R. (1938a). 'Some observations on the biology of the trout (*Salmo trutta*) in Windermere.' *J. Anim. Ecol.* **7**, 333–49.

ALLEN, K. R. (1938b). 'Deterioration of Windermere Trout.' *Salm. Trout Mag.* 1–7.

ALLEN, K. R. (1939). 'A note on the food of pike (*Esox lucius*) in Windermere.' *J. Anim. Ecol.* **8**, 72–5.

BAGENAL, T. B. (1966). 'The Ullswater Schelly.' *Field Naturalist*, **11**, 18–20.

BEAUCHAMP, R. S. A. (1932). 'Some ecological factors and their influence on competition between stream and lake-living triclads.' *J. Anim. Ecol.* **1**, 175–90.

BOYCOTT, A. E. (1936). 'The habitats of fresh-water Mollusca in Britain.' *J. Anim. Ecol.* **5**, 116–86.

BRINKHURST, R. O. (1963). 'A guide for the identification of British aquatic Oligochaeta.' *Sci. Publ. Freshwat. biol. Ass.* No. 22, pp. 52.

BRINKHURST, R. O. (1964). 'Observations on the biology of lake-dwelling Tubificidae.' *Arch. Hydrobiol.* **60**, 385–418.

CANNON, D., LUND, J. W. G. and SIEMINSKA, J. (1961). 'The growth of *Tabellaria flocculosa* (Roth) Kütz. var. *flocculosa* (Roth) Knuds. under natural conditions of light and temperature.' *J. Ecol.* **49**, 277–87.

CANTER, H. M. and LUND, J. W. G. (1948). 'Studies on plankton parasites. 1. Fluctuations in the numbers of *Asterionella formosa* Hass. in relation to fungal epidemics.' *New Phytol.* **47**, 238–61.

CANTER, H. M. and LUND, J. W. G. (1951). 'Studies on plankton parasites. 3. Examples of the interaction between parasitism and other factors determining the growth of diatoms.' *Ann. Bot. Lond.* N.S. **15**, 359–71.

CANTER, H. M. and LUND, J. W. G. (1953). 'Studies on plankton parasites. 2. The parasitism of diatoms, with special reference to lakes in the English Lake District.' *Trans. Br. mycol. Soc.* **36**, 13–37.

CANTER, H. M. and LUND, J. W. G., (1968). 'The importance of Protozoa in controlling the abundance of planktonic algae in lakes.' *Proc. Linn. Soc. Lond.* **179,** 203–19.

CLARIDGE, M. F. and STADDON, B. W. (1961). '*Stenelmis canaliculata* Gyll. (Col. Elmidae) a species new to the British list.' *Ent. mon. Mag.* **96,** 141–4.

COLEBROOK, J. M. (1960). 'Plankton and water movements in Windermere.' *J. Anim. Ecol.* **29,** 217–40.

COLLINGWOOD, R. G. (1933). 'An introduction to the prehistory of Cumberland, Westmorland and Lancashire north of the sands.' *Trans. Cumb. Westm. Antiq. Archaeol. Soc.* **33** (N.S.), 163–200.

COLLINGWOOD, W. G. (1925). *Lake District History*. Kendal: Titus Wilson, viii + 175.

COLLINS, V. G. (1957). 'Plankton bacteria.' *J. gen. Microbiol.* **16,** 268–72.

COLLINS, V. G. (1961). 'The distribution and ecology of gram-negative organisms other than Enterobacteriaceae in lakes.' *J. appl. Bact.* **23,** 510–14, 1960.

COLLINS, V. G. (1963). 'The distribution and ecology of bacteria in freshwater.' *Proc. Soc. Wat. Treat. Exam.*, 12, 40–73.

COLLINS, V. G. and KIPLING, C. (1957). 'The enumeration of waterborne bacteria by a new direct count method.' *J. appl. Bact.* **20,** 257–64.

COLLINS, V. G. and WILLOUGHBY, L. G. (1962). 'The distribution of bacteria and fungal spores in Blelham Tarn with particular reference to an experimental overturn.' *Arch. Mikrobiol.* **43,** 294–307.

DEWDNEY, J. C., TAYLOR, S. A. and WARDHAUGH, K. G. (1959). 'The Langdales: a lakeland parish.' *Occ. Pap. Dep. Geogr. Durham Univ.* No. 3, 23 pp.

DOTTRENS, E. (1959). 'Sur les corégones de Grande-Bretagne et d'Irlande.' *Int. Cong. Zool.* **15,** 404–6.

EDINGTON, J. M. (1964). 'The taxonomy of British polycentropid larvae (*Trichoptera*).' *Proc. zool. Soc. Lond.* **143** (2), 281–300.

ELLISON, N. F. (1966). 'Notes on the lakeland Schelly (*Coregonus clupeoides stigmaticus,* Regan).' *The Changing Scene*, No. 3, 46–53.

FRANKS, J. W. and PENNINGTON, W. (MRS T. G. TUTIN) (1961). 'The late-glacial and post-glacial deposits of the Esthwaite Basin, North Lancashire.' *New Phytol.* **60,** 27–42.

FROST, W. E. (1943). 'The natural history of the minnow *Phoxinus phoxinus*.' *J. Anim. Ecol.* **12,** 139–62.

FROST, W. E. (1945). 'The age and growth of eels (*Anguilla anguilla*) from the Windermere catchment area.' Part II, *J. Anim. Ecol.* **14,** 106–24.

FROST, W. E. (1946a). 'On the food relationships of fish in Windermere.' *Biol. Jaar. Dodonaea* **13,** 216–31.

FROST, W. E. (1946b). 'Observations on the food of eels (*Anguilla anguilla*) from the Windermere catchment area.' *J. Anim. Ecol.* **15,** 43–53.

FROST, W. E. (1954). 'The food of pike, *Esox lucius* L., in Windermere.' *J. Anim. Ecol.* **23,** 339–60.

FROST, W. E. (1956). 'The growth of brown trout (*Salmo trutta* L.) in Haweswater before and after the raising of the level of the lake.' *Salm. Trout. Mag.* **148,** 266–74.

FROST, W. E. (1965). 'Breeding habits of Windermere charr, *Salvelinus willughbii* (Günther), and their bearing on speciation of these fish.' *Proc. roy. Soc.* (*B*) **163,** 232–84.

FROST, W. E. and BROWN, M. E. (1967). *The Trout* (New Naturalist Special Volume). London: Collins, pp. 286.

FROST, W. E. and KIPLING, C. (1961). 'Some observations on the growth of the pike, *Esox lucius* in Windermere.' *Verh. int. Ver. Limnol.* **14,** 776–81.

FROST, W. E. and KIPLING, C. (1967a). 'A study of reproduction, early life, weight-length relationship and growth of pike, *Esox lucius* L., in Windermere.' *J. Anim. Ecol.* **36,** 651–93.

FROST, W. E. and KIPLING, C. (1967b). 'Removal of pike (*Esox lucius*) from Windermere and some of its effects on the population dynamics of that fish.' *Proc. 3rd Brit. Coarse Fish. Conf.* 53–6.

FROST, W. E. and MACAN, T. T. (1948). 'Corixidae (Hemiptera) as food of fish.' *J. Anim. Ecol.* **17,** 174–9.

FRYER, G. (1957). 'The food of some freshwater cyclopoid copepods and its ecological significance.' *J. Anim. Ecol.* **26,** 263–86.

GALLIFORD, A. L. (1947). 'Rotifera Report No. 1 for 1947.' *Lancs. Cheshire Fauna Cttee. Rep. No. 27,* 55–70.

GALLIFORD, A. L. (1949). 'Rotifera of Lancashire and Cheshire Report No. 2.' *Lancs. Cheshire Fauna Cttee. Report 1948–49,* 108–14.

GODWARD, M. B. (1937). 'An ecological and taxonomic investigation of the littoral algal flora of Lake Windermere.' *J. Ecol.* **25,** 496–568.

GODWIN, H. (1961). 'Radiocarbon dating and quaternary history in Britain.' *Proc. roy. Soc.* (*B*), **153,** 287–320.

GORHAM, E. (1955). 'On the acidity and salinity of rain.' *Geochim. et cosmochim Acta,* **7,** 231–9.

GORHAM, E. (1958). 'The influence and importance of daily weather conditions in the supply of chloride, sulphate and other ions to fresh waters from atmospheric precipitation.' *Phil. Trans.* (B) **241**, 147–78.

GOULDEN, C. E. 1964. 'The history of the Cladoceran fauna of Esthwaite Water (England) and its limnological significance.' *Arch. Hydrobiol.* **60**, 1–52.

HARDING, J. P. and SMITH, W. A. (1960). 'A key to the British freshwater cyclopid and calanoid copepods.' *Sci. Publ. Freshwat. biol. Ass.* No. 18, pp. 54.

HERON, J. (1961). 'The seasonal variation of phosphate, silicate, and nitrate in waters of the English Lake District.' *Limnol. Oceanogr.* **6**, 338–46.

HERON, J. and MACKERETH, F. J. H. (1955). 'The estimation of calcium and magnesium in natural waters, with particular reference to those of low alkalinity.' *Mitt. int. Ver. Limnol.* **5**, pp. 7.

HUGHES, J. C. and LUND, J. W. G. (1962). 'The rate of growth of *Asterionella formosa* Hass. in relation to its ecology.' *Arch. Mikrobiol.* **42**, 117–29.

HUMPHRIES, C. F. (1936). 'An investigation of the profundal and sub-littoral fauna of Windermere.' *J. Anim. Ecol.* **5**, 29–52.

HYNES, H. B. N. (1958). 'A key to the adults and nymphs of British stoneflies (Plecoptera).' *Sci. Publ. Freshwat. biol. Ass.* No. 17, pp. 87.

HYNES, H. B. N. (1960). *The biology of polluted waters*. Liverpool U.P., xiv + 202.

JENKIN, P. M. (1942). 'Seasonal changes in the temperature of Windermere (English Lake District).' *J. Anim. Ecol.* **11**, 248–69.

JENKIN, B. M. and MORTIMER, C. H. (1938). 'Sampling lake deposits.' *Nature, Lond.* **142**, 834.

JENKIN, B. M., MORTIMER, C. H. and PENNINGTON, W. (1941). 'The study of lake deposits.' *Nature, Lond.* **147**, 496–500.

KIMMINS, D. E. (1954). 'A revised key to the adults of the British species of Ephemeroptera.' *Sci. Publ. Freshwat. biol. Ass.* No. 15, pp. 71.

KIPLING, C. (1961). 'A salt spring in Borrowdale.' *Trans. Cumb. Westm. antiq. archeol. Soc.*, N.S. **61**, 57–70.

KNUDSON, B. M. (1954). 'The ecology of the diatom genus *Tabellaria* in the English Lake District.' *J. Ecol.* **42**, 345–58.

KNUDSON, B. M. (MRS T. H. KIPLING) (1957). 'Ecology of the epiphytic diatom *Tabellaria flocculosa* (Roth) Kütz. var. *flocculosa* in three English lakes.' *J. Ecol.* **45**, 93–112.

References

LE CREN, E. D. (1954). 'A subcutaneous tag for fish.' *J. Cons. int. Explor. Mer.* **20,** 72–82.

LE CREN, E. D. (1955). 'Year to year variations in the year-class strength of *Perca fluviatilis.*' *Verh. int. Ver. Limnol.* **12,** 187–92.

LE CREN, E. D. (1958). 'Observations on the growth of perch (*Perca fluviatilis* L.) over twenty-two years with special reference to the effects of temperature and changes in population density.' *J. Anim. Ecol.* **27,** 287–334.

LE CREN, E. D. (1959). 'Experiments with populations of fish in Windermere.' *Advanc. Sci., Lond.* **15,** 534–8.

LE CREN, E. D. (1961). 'How many fish survive?' *Yb. River Bds Ass.* **9,** 57–64.

LE CREN, E. D. (1965). 'Some factors regulating the size of populations of freshwater fish.' *Mitt. int. Ver. Limnol.* **13,** 88–105.

LUND, J. W. G. (1949a). 'Studies on *Asterionella.* 1. The origin and nature of the cells producing seasonal maxima.' *J. Ecol.* **37,** 389–419.

LUND, J. W. G. (1949b). 'The dynamics of diatom outbursts, with special reference to *Asterionella.*' *Verh. int. Ver. Limnol.* **10,** 275–6.

LUND, J. W. G. (1950). 'Studies on *Asterionella formosa* Hass. II. Nutrient depletion and the spring maximum.' *J. Ecol.* **38,** 1–35.

LUND, J. W. G. (1954). 'The seasonal cycle of the plankton diatom, *Melosira italica* (Ehr.) Kütz. subsp. *subarctica* O. Müll.' *J. Ecol.* **42,** 151–79.

LUND, J. W. G. (1955). 'Further observations on the seasonal cycle of *Melosira italica* (Ehr) Kütz. subsp. subarctica O. Müll.' *J. Ecol.* **43,** 90–102.

LUND, J. W. G. (1957). 'Chemical analysis in ecology illustrated from Lake District tarns and lakes. 2. Algal differences.' *Proc. Linn. Soc. Lond.* **167,** 165–71.

LUND, J. W. G. (1959a). 'A simple counting chamber for nannoplankton.' *Limnol. Oceanogr.* **4,** 57–65.

LUND, J. W. G. (1959b). 'Buoyancy in relation to the ecology of the freshwater phytoplankton.' *Brit. phycol. Bull.* No. 7, 1–17.

LUND, J. W. G. (1961). 'The periodicity of μ-algae in three English lakes.' *Verh. int. Ver. Limnol.* **14,** 147–54.

LUND, J. W. G. (1962). 'Concerning a counting chamber for nannoplankton described previously.' *Limnol. Oceanogr.* **7,** 261–2.

LUND, J. W. G. (1964). 'Primary production and periodicity of phytoplankton.' *Verh. int. Ver. Limnol.* **15,** 37–56.

LUND, J. W. G. (1965). 'The ecology of the freshwater phytoplankton.' *Biol. Rev.* **40**, 231–93.

LUND, J. W. G., KIPLING, C. and LE CREN, E. D. (1958). 'The inverted microscope method of estimating algal numbers and the statistical basis of estimations by counting.' *Hydrobiologia*, **11**, 143–70.

LUND, J. W. G., MACKERETH, F. J. H. and MORTIMER, C. H. (1963). 'Changes in depth and time of certain chemical and physical conditions and of the standing crop of *Asterionella formosa* Hass. in the North Basin of Windermere in 1947.' *Phil. Trans.* (B), **246**, 731, 255–290.

LUND, J. W. G. and TALLING, J. F. (1957). 'Botanical limnological methods with special reference to the algae.' *Bot. Rev.* **23**, 489–583.

MACAN, T. T. (1938). 'Evolution of aquatic habitats with special reference to the distribution of Corixidae.' *J. Anim. Ecol.* **7**, 1–19.

MACAN, T. T. (1940). 'Dytiscidae and Haliplidae (Col.) in the Lake District.' *Trans. Soc. Brit. Ent.* **7**, 1–20.

MACAN, T. T. (1949). 'Corixidae (Hemiptera) of an evolved lake in the English Lake District.' *Hydrobiologia*, **2**, 1–23.

MACAN, T. T. (1950). 'Ecology of fresh-water Mollusca in the English Lake District.' *J. Anim. Ecol.* **19**, 124–46.

MACAN, T. T. (1954a). 'The Corixidae (Hemipt.) of some Danish lakes.' *Hydrobiologia*, **6**, 44–69.

MACAN, T. T. (1954b). 'A contribution to the study of the ecology of Corixidae (Hemipt.).' *J. Anim. Ecol.* **23**, 115–41.

MACAN, T. T. (1955). 'Littoral fauna and lake types.' *Verh. int. Ver. Limnol.* **12**, 608–12.

MACAN, T. T. (1958). 'Methods of sampling the bottom fauna in stony streams.' *Mitt. int. Ver. Limnol.* **8**, pp. 21.

MACAN, T. T. (1961a). 'A key to the nymphs of the British species of Ephemeroptera.' *Sci. Publ. Freshwat. biol. Ass.* No. 20, pp. 64.

MACAN, T. T. (1961b). 'A review of running water studies.' *Verh. int. Ver. Limnol.* **14**, 587–602.

MACAN, T. T. (1962a). 'Why do some pieces of water have more Corixidae than others?' *Arch. Hydrobiol.* **58**, 224–32.

MACAN, T. T. (1962b). 'Biotic factors in running water.' *Schweiz. Z. Hydrol.* **24**, 386–407.

MACAN, T. T. (1963). *Freshwater Ecology.* London: Longmans, Green, x + 338.

MACAN, T. T. (1964). 'The Odonata of a moorland fishpond.' *Int. Rev. ges. Hydrobiol.* **49**, 325–60.

References

MACAN, T. T. (1965a). 'Predation as a factor in the ecology of water bugs.' *J. Anim. Ecol.* **34,** 691–8.

MACAN, T. T. (1965b). 'The influence of predation on the composition of fresh-water communities.' *Biological problems in water pollution.* Third seminar 1962. Cincinnati. 141–4.

MACAN, T. T. (1967). 'The Corixidae of two Shropshire meres.' *Field Studies,* **2,** 533–5.

MACAN, T. T. and MAUDSLEY, R. (1968). 'The insects of the stony substratum of Windermere.' *Trans. Soc. Brit. Ent.* **18,** 1–18.

MACAN, T. T., MCCORMACK, J. C. and MAUDSLEY, R. (1966/7). 'An experiment with trout in a moorland fishpond.' *Salm. Trout Mag.* **178,** 206–11; **179,** 59–69.

MACAN, T. T. and WORTHINGTON, E. B. (1951). *Life in lakes and rivers.* London: Collins, xvi + 272.

MACKERETH, F. J. H. (1951). 'A colorimetric method for routine estimation of calcium in natural waters.' *Analyst,* **76,** 482–4.

MACKERETH, F. J. H. (1953). 'Phosphorus utilization by *Asterionella formosa* Hass.' *J. Exp. Bot.,* **4,** 296–313.

MACKERETH, F. J. H. (1957). 'Chemical analysis in ecology illustrated from Lake District tarns and lakes. 1. Chemical analysis.' *Proc. Linn. Soc. Lond.* **167,** 159–64.

MACKERETH, F. J. H. (1958). 'A portable core sampler for lake deposits.' *Limnol. Oceanogr.* **3,** 181–91.

MACKERETH, F. J. H. (1963). 'Some methods of water analysis for limnologists.' *Sci. Publ. Freshwat. biol. Ass.* No. 21, pp. 71.

MACKERETH, F. J. H. (1964). 'An improved galvanic cell for determination of oxygen concentration in fluids.' *J. Sci. Instrum.* **41,** 38–41.

MACKERETH, F. J. H. (1965). 'Chemical investigation of lake sediments and their interpretation.' *Proc. roy. Soc.* (B). **161,** 295–309.

MACKERETH, F. J. H. (1966). 'Some chemical observations on post-glacial lake sediments.' *Phil. Trans.* (B), **250** (765), 165–213.

MACKERETH, J. C. (1954). 'Taxonomy of the larvae of the British species of the genus *Rhyacophila* (Trichoptera).' *Proc. R. ent. Soc. Lond.* (A) **29,** 147–52.

MACKERETH, J. C. (1956). 'Taxonomy of the larvae of the British species of the sub-family *Glossosomatinae* (Trichoptera).' *Proc. R. ent. Soc. Lond.* **31** (10–12), 167–72.

MAITLAND, P. S. (1966). 'Present status of known populations of the Vendace, *Coregonus vandesius,* Richardson, in Great Britain.' *Nature, Lond.* **210,** 216–17.

MANN, K. H. (1964). 'A key to the British freshwater leeches with notes on their ecology.' *Sci. Publ. Freshwat. biol. Ass.* No. 14, pp. 50.

MARR, J. E. (1916). *The geology of the Lake District and the scenery as influenced by geological structure.* Cambridge U.P., pp. 220.

MCCLEAN, W. N. (1940). 'Windermere basin: rainfall, run-off and storage.' *Quart. J.R. Met. Soc.* **66**, 337–62.

MCCORMACK, J. C. (1965). 'Observations on the perch population of Ullswater.' *J. Anim. Ecol.* **34**, 463–78.

MCCORMACK, J. C. (in press). 'Some observations on the food of perch (*Perca fluviatilis* L.) in Windermere.' *J. Anim. Ecol.*

MILL, H. R. (1895). 'Bathymetrical survey of the English Lakes.' *Geogr. J.* **6**, 46–73, 135–66.

MISRA, R. D. (1938). 'Edaphic factors in the distribution of aquatic plants in the English Lakes.' *J. Ecol.* **26**, 411–51.

MONKHOUSE, F. J. (1960). *The English Lake District. British landscapes through maps.* 1. Sheffield: Geogr. Assoc., pp. 19.

MOON, H. P. (1934). 'An investigation of the littoral region of Windermere.' *J. Anim. Ecol.* **3**, 8–28.

MOON, H. P. (1935a). 'Methods and apparatus suitable for an investigation of the littoral region of oligotrophic lakes.' *Int. Rev. Hydrobiol.* **32**, 319–33.

MOON, H. P. (1935b). 'Flood movements of the littoral fauna of Windermere.' *J. Anim. Ecol.* **4**, 216–28.

MOON, H. P. (1936). 'The shallow littoral region of a bay at the north-west end of Windermere.' *Proc. Zool. Soc. Lond.* 490–515.

MOON, H. P. (1956). 'The distribution of *Potamopyrgus jenkinsi* (E. A. Smith) in Windermere.' *Proc. Malacol. Soc.* **32**, 105–8.

MOON, H. P. (1957). 'The distribution of *Asellus* in Windermere.' *J. Anim. Ecol.* **26**, 113–23.

MOON, H. P. (1968). 'The colonization of Esthwaite Water and Ullswater, English Lake District, by *Asellus* (Crustacea, Isopoda).' *J. Anim. Ecol.* **37**, 405–15.

MOORE, W. H. (1954). 'A new type of electrical fish-catcher.' *J. Anim. Ecol.* **23**, 373–5.

MOORE, W. H. and MORTIMER, C. H. (1954). 'A portable instrument for the location of subcutaneous fish-tags.' *J. Cons. int. Explor. Mer.* **20**, 83–6.

MORTIMER, C. H. (1938). 'The nitrogen balance of large bodies of water.' *Public Health Services Congress 1938*, paper no. 16a, 3–11.

References

MORTIMER, C. H. (1939). 'Physical and chemical aspects of organic production in lakes.' *Proc. Assoc. appl. Biol.* **26,** 167–72.

MORTIMER, C. H. (1940). 'An apparatus for obtaining water from different depths for bacteriological examination.' *J. Hyg. Camb.* **40,** 641–6.

MORTIMER, C. H. (1941/42). 'The exchange of dissolved substances between mud and water in lakes.' I and II. *J. Ecol.* **29,** 280–329. III and IV. *J. Ecol.* **30,** 147–201.

MORTIMER, C. H. (1952). 'Water movements in lakes during summer stratification; evidence from the distribution of temperature in Windermere.' *Phil. Trans.* (B) **236,** 355–404.

MORTIMER, C. H. (1953). 'A review of temperature measurement in limnology.' *Mitt. int. Ver. Limnol.* **1,** pp. 25.

MORTIMER, C. H. (1955). 'The dynamics of the autumn overturn in a lake.' *Gen. Ass. int. Geod., Rome,* **3,** 15–24.

MORTIMER, C. H. (1959). 'The physical and chemical work of the Freshwater Biological Association, 1935–57.' *Advanc. Sci., Lond.,* **61,** 524–30.

MORTIMER, C. H. and MOORE, W. H. (1953). 'The use of thermistors for the measurement of lake temperatures.' *Mitt. int. Ver. Limnol.* **2,** pp. 42.

MORTIMER, C. H. and WORTHINGTON, E. B. (1942). 'Morphometric data for Windermere.' *J. Anim. Ecol.* **11,** 245–7.

MUNDIE, J. H. (1956). 'Emergence traps for aquatic insects.' *Mitt. int. Ver. Limnol.* **7,** pp. 13.

MUNDIE, J. H. (1964). 'The activity of benthic insects in the water mass of lakes.' *Proc. int. Congr. Ent.* **12,** 410.

PEARSALL, W. H. (1917). 'The aquatic and marsh vegetation of Esthwaite Water.' *J. Ecol.* **5,** 180–202.

PEARSALL, W. H. (1920). 'The aquatic vegetation of the English Lakes.' *J. Ecol.* **8,** 163–201.

PEARSALL, W. H. (1921). 'The development of vegetation in the English Lakes, considered in relation to the general evolution of glacial lakes and rock basins.' *Proc. roy. Soc.* (B) **92,** 259–84.

PEARSALL, W. H. (1924). 'Phytoplankton and environment in the English Lake District.' *Rev. Algol.* 1–15.

PEARSALL, W. H. (1932). 'Phytoplankton in the English Lakes. II. The composition of the phytoplankton in relation to dissolved substances.' *J. Ecol.* **20,** 241–62.

PEARSALL, W. H. (1959). 'The Freshwater Biological Association and its laboratory.' *Advanc. Sci., Lond.* **61,** 521–3.

PEARSALL, W. H. and HEWITT, T. (1933). 'Light penetration into fresh water. 2. Light penetration and changes in vegetation limits in Windermere.' *J. Exp. Biol.* **10**, 306–12.

PEARSALL, W. H. and PEARSALL, W. H. (1925). 'Phytoplankton of the English Lakes.' *J. Linn. Soc. Bot.* **47**, 55–73.

PEARSALL, W. H. and PENNINGTON, W. (MRS T. G. TUTIN) (1947). 'Ecological history of the English Lake District.' *J. Ecol.* **34**, 137–48.

PEARSALL, W. H. and ULLYOTT, P. (1934). 'Light penetration into fresh water. 3. Seasonal variations in the light conditions in Windermere in relation to vegetation.' *J. exp. Biol.* **11**, 89–93.

PENNINGTON, W. (MRS T. G. TUTIN) (1943). 'Lake sediments: the bottom deposits of the North Basin of Windermere, with special reference to the diatom succession.' *New Phytol.* **42**, 1–27.

PENNINGTON, W. (MRS T. G. TUTIN) (1947). 'Studies of the post-glacial history of British vegetation. VII. Lake sediments: pollen diagrams from the bottom deposits of the North basin of Windermere.' *Phil. Trans.* (B) **233**, 137–75.

PENNINGTON, W. (MRS T. G. TUTIN) (1964). 'Pollen analysis from the deposits of six upland tarns in the Lake District.' *Phil. Trans.* (B) **248**, 746, 205–44.

PENNINGTON, W. (MRS T. G. TUTIN) (1965). 'The interpretation of some post-glacial vegetation diversities at different Lake District sites.' *Proc. roy. Soc.* (B) **161**, 310–23.

PENNINGTON, W. (MRS T. G. TUTIN) and FROST, W. E. (1961). 'Fish vertebrae and scales in a sediment core from Esthwaite Water (English Lake District).' *Hydrobiologia*, **17**, 183–90.

READER'S DIGEST 1965. *Complete Atlas of the British Isles.* London: Reader's Digest Assoc., pp. 230.

REGAN, C. T. (1908). 'A revision of the British and Irish fishes of the genus *Coregonus*.' *Ann. Mag. Nat. Hist.* ser. 8. **2**, 482–90.

REGAN, C. T. (1911). *The freshwater fishes of the British Isles.* London: Methuen, xxv + 287.

REYNOLDSON, T. B. (1966). 'The distribution and abundance of lake-dwelling triclads: towards a hypothesis.' *Advanc. Ecol. Res.* **3**, pp. 71.

REYNOLDSON, T. B. (1967). 'A key to the British species of freshwater triclads.' *Sci. Publ. Freshwat. biol. Ass.* No. 23, pp. 28.

ROLLINSON, W. (1967). *A history of man in the Lake District.* London: Dent, xii + 162.

ROUND, F. E. (1957a). 'The late-glacial and post-glacial diatom succession in the Kentmere Valley deposit.' *New Phytol.* **56**, 98–126.

ROUND, F. E. (1957b). 'The distribution of Bacillariophyceae on some littoral sediments of the English Lake District.' *Oikos*, **8**, 16–37.

ROUND, F. E. (1957c). 'Studies on bottom-living algae in some lakes of the English Lake District. 1. Some chemical features of the sediments related to algal productivities.' *J. Ecol.* **45**, 133–48.

ROUND, F. E. (1957d). 'Studies on bottom-living algae in some lakes of the English Lake District. II. The distribution of Bacillariophyceae on the sediments.' *J. Ecol.* **45,** 343–60.

ROUND, F. E. (1957e). 'Studies on bottom-living algae in some lakes of the English Lake District. III. The distribution on the sediments of algal groups other than the Bacillariophyceae.' *J. Ecol.* **45**, 649–64.

ROUND, F. E. (1960). 'Studies on bottom-living algae in some lakes of the English Lake District. Part II. The seasonal cycles of the Bacillario-phyceae.' *J. Ecol.* **48**, 529–47.

ROUND, F. E. (1961a). 'Studies on bottom-living algae in some lakes of the English Lake District. Part V. The seasonal cycles of the Cyano-phyceae.' *J. Ecol.* **49**, 31–8.

ROUND, F. E. (1961b). 'Studies on the bottom-living algae in some lakes of the English Lake District. Part 6. The effect of depth on the epipelic algal community.' *J. Ecol.* **49**, 245–54.

ROUND, F. E. (1961c). 'The diatoms of a core from Esthwaite Water.' *New Phytol.* **60**, 43–59.

SCOURFIELD, D. J. and HARDING, J. P. (1958). 'A key to the British species of freshwater Cladocera.' *Sci. Publ. Freshwat. biol. Ass.* No. 5, pp. 55.

SMYLY, W. J. P. (1955). 'On the biology of the stone-loach, *Nemacheilus barbatula* (L.).' *J. Anim. Ecol.* **24**, 167–86.

SMYLY, W. J. P. (1957). 'The life history of the Bullhead or Miller's Thumb (Cottus gobio L.).' *Proc. zool. Soc. Lond.* **128**, 431–53.

SMYLY, W. J. P. (1961). 'The life-cycle of the freshwater copepod *Cyclops leuckarti* Claus in Esthwaite Water.' *J. Anim. Ecol.* **30**, 153–69.

SMYLY, W. J. P. (1968). 'Observations on the planktonic and profundal Crustacea of the lakes of the English Lake District.' *J. Anim. Ecol.* **37**, 693–708.

SWIFT, D. R. (1961). 'The annual growth-rate cycle in Brown Trout (*Salmo trutta* Linn.) and its cause.' *J. exp. Biol.* **38**, 595–604.

SWIFT, D. R. (1962). 'Activity cycles in the Brown Trout (*Salmo trutta* Lin.) 1. Fish feeding naturally.' *Hydrobiologia*, **20**, 241–7.

SWIFT, D. R. (1964). 'Activity cycles in the Brown Trout (*Salmo trutta* L.) 2. Fish artificially fed.' *J. Fish. Res. Bd Can.* **21**, 133–8.

SWYNNERTON, G. H. and WORTHINGTON, E. B. (1939). 'Brown-trout growth in the Lake District.' *Salm. Trout Mag.* No. 97, 337–55.

SWYNNERTON, G. H. and WORTHINGTON, E. B. (1940). 'Note on the food of fish in Haweswater (Westmorland).' *J. Anim. Ecol.* **9**, 183–7.

TALLING, J. F. (1957). 'The growth of two plankton diatoms in mixed cultures.' *Physiol. Plant.* **10**, 215–23.

TALLING, J. F. (1961). 'Photosynthesis under natural conditions.' *Annu. Rev. Pl. Physiol.* **12**, 133–54.

TALLING, J. F. (1966). 'Photosynthetic behaviour in stratified and unstratified lake populations of a planktonic diatom.' *J. Ecol.* **54**, 99–127.

TANSLEY, A. (1939). *The British Islands and their vegetation.* C.U.P., xxxviii + 930.

TAYLOR, C. B. (1939). 'Bacteria of lakes and impounded waters.' *Off. Circ. Brit. Waterw. Ass.* **21**, 616–23.

TAYLOR, C. B. (1940). 'Bacteriology of fresh water. I. Distribution of bacteria in English lakes.' *J. Hyg. Camb.* **40**, 616–40.

TAYLOR, C. B. (1941a). 'Bacteriology of fresh water. II. The distribution and types of coliform bacteria in lakes and streams.' *J. Hyg. Camb.* **41**, 17–38.

TAYLOR, C. B. (1941b). 'The distribution of Bacteria in lakes and their inflows.' *Proc. Soc. agric. Bact.* 1941, 1–6.

TAYLOR, C. B. (1942a). 'Bacteriology of freshwater. III. The types of bacteria present in lakes and streams and their relationship to the bacterial flora of soil.' *J. Hyg. Camb.* **42**, 284–96.

TAYLOR, C. B. (1942b). 'The ecology and significance of the different types of coliform bacteria found in water. A review of the literature.' *J. Hyg. Camb.* **42**, 23–44.

TAYLOR, C. B. (1948). 'The bacteriology of lakes.' *Endeavour*, **7**, 111–15.

TAYLOR, C. B. (1949a). 'The effect of phosphorus on the decomposition of organic matter in fresh water.' *Proc. Soc. appl. Bact.* 1949, No. 2, 96–9.

TAYLOR, C. B. (1949b). 'Factors affecting the bacterial population of lake waters.' *Proc. Soc. appl. Bact.* **1**, 4–10.

TUTIN, W. (PENNINGTON, W.) (1955). 'Preliminary observations on a year's cycle of sedimentation in Windermere, England.' *Mem. Ist. ital. Idrobiol. de Marchi*, Suppl. 8, 467–84.

References

ULLYOTT, P. (1937). 'An apparatus for plankton counting.' *Int. Rev. Hydrobiol.* **34**, 15–23.

ULLYOTT, P. (1939). 'Die täglichen Wanderungen der planktonischen Süsswasser-Crustaceen.' *Int. Rev. Hydrobiol.* **38**, 262–84.

WAILES, G. H. (1939). 'The plankton of Lake Windermere, England.' *Ann. Mag. Nat. Hist.* (Ser. II) **3**, 401–14.

WATSON, J. (1925). *The English Lake District fisheries.* London & Edinburgh: Foulis, xv + 271.

WEBB, M. G. (Mrs D. M. Guthrie) (1961). 'The effects of thermal stratification on the distribution of benthic Protozoa in Esthwaite Water.' *J. Anim. Ecol.* **30**, 137–51.

WEST, W. and WEST, G. S. (1909). 'The phytoplankton of the English Lake District.' *The Naturalist* No. 626, 115–22; 627, 134–41; 628, 186–93; 630, 260–7; 631, 287–92; 632, 323–31.

WILLOUGHBY, L. G. (1962). 'The ecology of some lower fungi in the English Lake District.' *Trans. Brit. mycol. Soc.* **45**, 121–36.

WILLOUGHBY, L. G. (1965). 'Some observations on the location of sites of fungal activity at Blelham Tarn.' *Hydrobiologia*, **25**, 352–6.

WILLOUGHBY, L. G. and COLLINS, V. G. (1966). 'A study of the distribution of fungal spores and Bacteria in Blelham Tarn and its associated streams.' *Nova Hedwigia*, **12**, 149–71.

WORTHINGTON, E. B. (1950). 'An experiment with populations of fish in Windermere, 1939–48.' *Proc. zool. Soc. Lond.* **120**, 113–49.

2. Other references cited

ALLEN, K. R. (1952). 'A New Zealand trout stream—some facts and figures.' *New Zealand Marine Dept. Fish. Bull.* **10**A, pp. 70.

ALM, G. (1952). 'Year class fluctuations and span of life of perch.' *Rep. Inst. Freshw. Res. Drottning.* No. 33, 17–38.

BERG, K. (1938). 'Studies on the bottom animals of Esrom Lake.' *K. danske vidensk. Selsk Skr.* **7**, pp. 255.

CARPENTER, K. E. (1926). 'The lead mine as an active agent in river pollution.' *Ann. appl. Biol.* **13**, 395–401.

CRAWFORD, G. I. (1937). 'An amphipod, *Eucrangonyx gracilis* S. I. Smith, new to Britain.' *Nature, Lond.* **139**, 327.

ELSTER, H.–J. (1958). 'Das limnologische Seetypensystem, Rückblick und Ausblick.' *Verh. int. Ver. Limnol.* **13**, 101–20.

FELIKSIAK, S. (1939). 'Physa acuta Draparnaud in den Fabrikteichen von Łódź und ihre allgemeine Verbreitung.' *Frag. Faun. Mus. Zool. Polonici* **4**, 243–58.

FINDENEGG, I. (1953). 'Kärntner Seen naturkundlich betrachtet. *Carinthia II.*' Sonderheft **15**, pp. 101.

HICKIN, N. E. (1967). *Caddis larvae. Larvae of the British Trichoptera.* London: Hutchinson, xi + 476.

HOLMQUIST, C. (1959). *Problems on marine-glacial relicts. An account of investigations of the genus Mysis.* Lund. Berling, pp. 270.

HYNES, H. B. N. (1955). 'Distribution of some freshwater Amphipoda in Britain.' *Verh. int. Ver. Limnol.* **12**, 620–8.

HYNES, H. B. N. (1961). 'The invertebrate fauna of a Welsh mountain stream.' *Arch. Hydrobiol.* **57**, 344–88.

JÓNASSON, P. M. (1958). 'The mesh factor in sieving techniques.' *Verh. int. Ver. Limnol.* **13**, 860–6.

MARLIER, G. (1951). 'La biologie d'un ruisseau de plaine.' *Mém. Inst. Sci. nat. Belg.* **114**, 1–98.

NIELSEN, A. (1950). 'On the zoogeography of springs.' *Hydrobiologia*, **2**, 313–21.

NILSSON, N.-A. (1963). 'Interaction between trout and char in Scandinavia.' *Trans. Amer. Fish. Soc.* **92**, 276–85.

NILSSON, N.-A. (1965). 'Food segregation between salmonoid species in North Sweden.' *Rep. Inst. Freshwat. Res. Drottningholm.* No. 46, 58–73.

PATTEE, E. (1965). 'Sténothermie et eurythermie. Les invertébrés d'eau douce et la variation journalière de température.' *Ann. Limnol.* **1**, 281–434.

RODHE, W. (1958). 'The primary production in lakes: some results and restrictions of the ^{14}C method.' *Rapp. Cons. Eplor. Mer.* **144**, 122–8.

ROSE, S. M. and ROSE, F. C. (1965). 'The control of growth and reproduction in freshwater organisms by specific products.' *Mitt. int. Ver. Limnol.* **13**, 21–35.

SLACK, H. D. (1957). *Studies on Loch Lomond.* 1. Glasgow U.P., x + 133.

SVÄRDSON, G. (1957). 'The Coregonid problem. VI. The palearctic species and their intergrades.' *Rep. Inst. Freshwat. Res. Drottningholm.* **38**, 267–356.

TATTERSALL, W. M. (1937). 'Occurrence of *Eucrangonyx gracilis.*' *Nature, Lond.* **139**, 593.

THIENEMANN, A. (1950). *Verbreitungsgeschichte der Süsswassertierwelt Europas.* Die Binnengewässer 18, xvi + 809. Stuttgart: Schweizerbart.

THIENEMANN, A. (1954). *Chironomus.* Die Binnengewässer 20, xvi + 834. Stuttgart: Schweizerbart.

References

VALVOVIRTA, E. J. (1958). 'Über die Bodentierbonitierungen im Päijänne —See in Finnland.' *Verh. int. Ver. Limnol.* **13**, 284–7.

WESENBERG-LUND, C. (1921). 'Contributions to the biology of the Danish Culicidae'. *K. Danske Vidensk. Selsk. Skr.* **7**, 1–210.

WILLIAMS, W. D. (1963). 'The ecological relationships of isopod crustaceans *Asellus aquaticus* (L.) and *A. meridianus* Rac.' *Proc. zool. Soc. Lond.* **140**, 4, 661–79.

Index to Persons

*R

243

Index of Scientific Names

Acanthocystis chaetophora, 98
  var. simplex, 100
  spinifera, 100
  tuffacea, 98
Achnanthes microcephala, 90
  minutissima, 90, 195, 197
Acroperus harpae, 201–3
Agapetus fuscipes, 142, 144, 145, 147–54
Alnus, 184
Alona
  affinis, 96, 201–3
  costata, 201, 203
  guttata, 96, 201
  intermedia, 201–3
  quadrangularis, 96, 201–3
  rectangula, 96, 201
Alonella
  excisa, 201–3
  exigua, 201–3
  nana, 96, 201–3
Alonopsis elongata, 201–3
Amphinemura sulcicollis, 158
Amphora gracilis, 92
Anabaena, 84, 101
  circinalis, 73, 74
  constricta, 92
  flos-aquae, 73, 74
  lemmermanni, 73, 74
Anarthra aptera, 99
Anchistropus emarginatus, 202, 203
Ancylus
  fluviatilis, 128, 129, 142–9, 151, 153, 158, 174
  lacustris, 128
Anguilla anguilla, 165
Anomoeoneis, 198
Aphanizomenon flos-aquae, 73, 84
Aphanothece stagnina, 93
Argonotholca foliacea, 99
Asellus, 149, 150, 152, 158, 159, 172, 173, 174, 216
  aquaticus, 126, 137, 142, 144, 145, 147, 151, 173
  meridianus, 137, 145, 151, 161
Asplanchna, 81, 83
  priodonta, 95, 97, 99
Asterionella, 63, 64, 69
  formosa, 70, 73–85, 196–8, 211
  gracillima (formosa), 73, 77, 78, 197
Aulodrilus pluriseta, 138

Aureobasidium pullulans, 212

Balanophrya mamillata, 100
Batrachospermum moniliforme, 91
Batracobdella paludosa, 145, 156
Bdellocephala punctata, 145, 155
Betula
  nana, 184
  pendula, 184
  pubescens, 184
Bithynia
  leachii, 129
  tentaculata, 129
Bosmina, 96, 97
  coregoni, 201, 204
  coregoni var. obtusirostris, 94, 98
  longirostris, 94, 98, 201, 204
Botryococcus braunii, 73, 74
Bulbochaete intermedia, 91
Bythotrephes longimanus, 94, 97, 98, 201

Caenis horaria, 160, 161
Caenomorpha medusula, 139
Callicorixa praeusta, 133–6
Callitriche, 105, 111, 131
  autumnalis, 111
  intermedia, 106, 108, 109, 114, 117
Calluna, 188
Caloneis
  amphisbaena, 92
  silicula, 92
Calothrix parietina, 90
Camptocercus rectirostris, 201–3
Campylodiscus noricus, 199, 200
Canthocamptus staphylinus, 96
Capnia bifrons, 143–5, 151
Carex
  elata, 132, 134
  rostrata, 110, 116, 118, 119, 132
Castalia
  alba, 115, 118
  minor, 107, 115, 118
Centroptilum luteolum, 142–5, 147, 151–3, 158
Cephalodela exigua, 99
Ceratium hirundinella, 73, 75, 85
Ceriodaphnia, 94, 96, 98
  megalops, 201
  pulchella, 201, 204
  quadrangula, 201, 204

245

Subject Index

Afforestation, 19, 20, 184
Agriculture
and production in lakes, 28, 196, 198
Algae, 72–93
attached, 87–93
counting, 215
dry weights of, 85
μ-, 86, 87
nuisance from, 223
phytoplankton in, 72–6
predators on, 81, 83
production by, 83–7
in productive lakes, 72–5
rate of sinking of, 82
spectographic analyses of, 70
succession of species of, 75, 76
temperature and rate of growth of, 81–3
unproductive lakes, in, 72–4, 76
Allerød, 183, 200
Aluminium, 119
Ambleside, 29, 41
development of, 20
meteorological station, 6
records at, 8–13
Roman Fort at, 17
sewage, 37, 141, 148, 149
Ammonia, 64, 119
Analyses, chemical
common elements, 59, 65–9
rare elements, 70
rainfall, 60, 61
substances that fluctuate, 65, 66
Archaeology, 16, 17
Asellus, spread of in Esthwaite, 126
in Windermere, 136, 137
Asterionella
competition with other algae, 79, 83
growth requirements, 64, 77, 81
parasitism of, 79
periodicity of, 77–81
phosphate uptake by, 63
predators on, 81
silica as limiting factor, 78
Axes, stone, 16, 188

Bacillariophyceae, 89
Bacteria, 206–12
actvities of, 211, 212
activity measured by O^2 consumption, 210
culture of, 206
direct counting of, 206
effect of dying algae on, 211
limiting factors of, 211
multiplication in lakes, 210
need for substrate of, 211
and rainfall, 208, 209
seasonal abundance of, 210
Bannisdale shore, 38
Bassenthwaite, 5, 15, 16, 30, 40, 65
attached algae in, 89
Coregonus in, 166
corixids in, 136, 137
Crangonyx in, 125
fish in, 166, 169
flatworms in, 155
gasteropods in, 129
leeches in, 156
location of, 5, 6, 29
morphometry, 31, 32
oligochaetes in, 138
oxygen profile of, 67
planktonic Crustacea in, 95–7
population of drainage area, 42
production of algae in, 86
rooted vegetation in, 116, 117
stone fauna of, 150–4
stratification in, 52
substances dissolved in, 69
temperature profile of, 67
transparency of, 56–8
Benthos (see communities)
Bicarbonate, 59, 69, 119
Birge–Ekman Grab, 216
Black Beck, 115, 116, 118
Blelham Tarn, 22, 40, 64, 65, 94
attached algae in, 89
bacteria in, 206–10
bacterial activity in, 211
corixids in, 132–4, 136, 137
destruction of stratification in, 206–10
gasteropods in, 128, 129
location of, 6
morphometry of, 31, 32
nitrogen balance of, 63
oxygen profile of, 67
planktonic Crustacea in, 94–7
production by algae in, 84–6

252

CAMROSE LUTHERAN COLLEGE
LIBRARY

QH
138
L 35
M 12
16,425